Bioremediation
through
Rhizosphere Technology

Bioremediation through Rhizosphere Technology

Todd A. Anderson, EDITOR
Iowa State University

Joel R. Coats, EDITOR
Iowa State University

Developed from a symposium sponsored
by the Division of Agrochemicals and
the Division of Environmental Chemistry
at the 206th National Meeting
of the American Chemical Society,
Chicago, Illinois
August 23–27, 1993

American Chemical Society, Washington, DC 1994

Library of Congress Cataloging-in-Publication Data

Bioremediation through rhizosphere technology / Todd A. Anderson, editor, Joel R. Coats, editor.

p.   cm.—(ACS symposium series, ISSN 0097–6156; 563)

"Developed from a symposium sponsored by the Division of Agrochemicals and the Division of Environmental Chemistry at the 206th National Meeting of the American Chemical Society, Chicago, Illinois, August 23–27, 1993."

Includes bibliographical references and indexes.

ISBN 0–8412–2942–2

1. Soil remediation—Congresses.   2. Bioremediation—Congresses. 3. Rhizosphere—Congresses.

I. Anderson, Todd A., 1963–   . II. Coats, Joel R. III. American Chemical Society. Division of Agrochemicals. IV. American Chemical Society. Division of Environmental Chemistry. V. American Chemical Society. Meeting (206th: 1993: Chicago, Ill.) VI. Series.

TD878.B557   1994
628.5′5—dc20   94–5331
                                                              CIP

The paper used in this publication meets the minimum requirements of American National Standard for Information Sciences—Permanence of Paper for Printed Library Materials, ANSI Z39.48–1984. ∞

Foreword

THE ACS SYMPOSIUM SERIES was first published in 1974 to provide a mechanism for publishing symposia quickly in book form. The purpose of this series is to publish comprehensive books developed from symposia, which are usually "snapshots in time" of the current research being done on a topic, plus some review material on the topic. For this reason, it is necessary that the papers be published as quickly as possible.

Before a symposium-based book is put under contract, the proposed table of contents is reviewed for appropriateness to the topic and for comprehensiveness of the collection. Some papers are excluded at this point, and others are added to round out the scope of the volume. In addition, a draft of each paper is peer-reviewed prior to final acceptance or rejection. This anonymous review process is supervised by the organizer(s) of the symposium, who become the editor(s) of the book. The authors then revise their papers according to the recommendations of both the reviewers and the editors, prepare camera-ready copy, and submit the final papers to the editors, who check that all necessary revisions have been made.

As a rule, only original research papers and original review papers are included in the volumes. Verbatim reproductions of previously published papers are not accepted.

M. Joan Comstock
Series Editor

Contents

Preface xi

RHIZOSPHERE TECHNOLOGY

1. Role of Microorganisms in Soil Bioremediation............................ 2
   Jean-Marc Bollag, Tawna Mertz, and Lewis Otjen

2. Toxicant Degradation in the Rhizosphere 11
   Barbara T. Walton, Elizabeth A. Guthrie, and
   Anne M. Hoylman

INTERACTIONS AMONG MICROORGANISMS, PLANTS, AND CHEMICALS

3. Polyphenolic Compounds Respired by Bacteria 28
   Andrei L. Barkovskii, Marie-Louise Boullant, and
   Jacques Balandreau

4. Effect of Nutrients on Interaction Between Pesticide
   Pentachlorophenol and Microorganisms in Soil............................ 43
   Kyo Sato

5. Intact Rhizosphere Microbial Communities Used To Study
   Microbial Biodegradation in Agricultural and Natural Soils:
   Influence of Soil Organic Matter on Mineralization
   Kinetics 56
   David B. Knaebel and J. Robie Vestal

6. Influence of Plant Species on In Situ Rhizosphere
   Degradation 70
   Jodi R. Shann and J. J. Boyle

7. Rhizosphere Microbial Communities as a Plant Defense
   Against Toxic Substances in Soils 82
   Barbara T. Walton, Anne M. Hoylman, Mary M. Perez,
   Todd A. Anderson, Theodore R. Johnson,
   Elizabeth A. Guthrie, and Russell F. Christman

8. Potential Use of Mycorrhizal Fungi as Bioremediation
   Agents.. 93
   Paula K. Donnelly and John S. Fletcher

INDUSTRIAL CHEMICALS

9. Dehalogenation of Chlorinated Phenols During Binding
   to Humus 102
   Jerzy Dec and Jean-Marc Bollag

10. Alfalfa Plants and Associated Microorganisms Promote
    Biodegradation Rather Than Volatilization of Organic
    Substances from Ground Water 112
    Lawrence C. Davis, N. Muralidharan, V. P. Visser,
    C. Chaffin, W. G. Fateley, L. E. Erickson, and
    R. M. Hammaker

11. Volatilization and Mineralization of Naphthalene
    in Soil–Grass Microcosms... 123
    J. W. Watkins, D. L. Sorensen, and R. C. Sims

12. Biologically Mediated Dissipation of Polyaromatic
    Hydrocarbons in the Root Zone.. 132
    A. P. Schwab and M. K. Banks

13. Grass-Enhanced Bioremediation for Clay Soils
    Contaminated with Polynuclear Aromatic Hydrocarbons............. 142
    X. Qiu, S. I. Shah, E. W. Kendall, D. L. Sorensen,
    R. C. Sims, and M. C. Engelke

PESTICIDES

14. Propanil Metabolism by Rhizosphere Microflora 160
    Robert E. Hoagland, Robert M. Zablotowicz, and
    Martin A. Locke

15. Glutathione S-Transferase Activity in Rhizosphere
    Bacteria and the Potential for Herbicide Detoxification 184
    Robert M. Zablotowicz, Robert E. Hoagland, and
    Martin A. Locke

16. Biological Degradation of Pesticide Wastes in the Root
    Zone of Soils Collected at an Agrochemical Dealership 199
    Todd A. Anderson, Ellen L. Kruger, and Joel R. Coats

17. **Plant and Microbial Establishment in Pesticide-Contaminated Soils Amended with Compost** 210
   Michael A. Cole, Xianzhong Liu, and Liu Zhang

INDEXES

Author Index 225

Affiliation Index 225

Subject Index 226

Preface

THE USE OF MICROORGANISMS to remediate environments contaminated by hazardous substances is an innovative technology and an area of intense interest. Although biological technology has been used for decades in wastewater treatment, recent examination of the cost-effectiveness of this technology has led to its application to hazardous chemicals at waste sites. Successes obtained by using the natural metabolic capabilities of bacteria and fungi to clean up soil, sediment, and water have encouraged continued interest and research in bioremediation.

Microbial degradation of toxicants can be hindered, when, for example, the population or activity of microorganisms capable of degrading the toxicants is limited by the environmental conditions. In such cases, environmental conditions must be altered to effectuate remediation. The use of vegetation to facilitate microbial degradation of toxicants may be a viable method for remediating contaminated environments in situ. The plant root zone, or rhizosphere, provides a habitat conducive to the proliferation of microbial growth and activity. Previous research indicated decreased persistence of pesticides in rhizosphere soils. Recent investigations revealed similar results with industrial chemicals such as surfactants, oil residues, polycyclic aromatic hydrocarbons (PAHs), pentachlorophenol, and trichloroethylene.

Although numerous texts deal with the rhizosphere, this is the first book on the potential use of the rhizosphere for bioremediation. The book is divided into four sections. Overviews of bioremediation and rhizosphere microbiology are provided in the first section. The interactions between microorganisms, plants, and chemicals in the rhizosphere are presented in six chapters contained in the second section. The degradation of industrial chemicals in the rhizosphere, including PAHs and chlorinated phenols, is presented in the third section. Finally, the section on microbial degradation of pesticides in the rhizosphere contains chapters on herbicides as well as applications to pesticide-contaminated sites. Overall, we believe the book provides the lay reader with valuable review information that puts in context the potential role of the rhizosphere in bioremediation. In addition, the book provides those active in this area of research with a document that summarizes the current state of the science.

Acknowledgments

We thank the contributors to this volume for providing information on the current knowledge of the influence of rhizosphere microorganisms in degrading pesticides and industrial chemicals. The technical reviewers for this volume deserve special recognition for their efforts in the peer-review process.

We express appreciation to the Agrochemicals and Environmental Chemistry Divisions of the American Chemical Society for providing assistance and a forum for this research. We also thank Anne Wilson of ACS Books for her assistance in bringing this book to fruition.

We dedicate this book to our wives, Brenda Anderson and Becky Coats, for their support during the development of this book and throughout our professional lives.

TODD A. ANDERSON
Pesticide Toxicology Laboratory
Department of Entomology
Iowa State University
Ames, IA 50011–3140

JOEL R. COATS
Pesticide Toxicology Laboratory
Department of Entomology
Iowa State University
Ames, IA 50011–3140

May 3, 1994

RHIZOSPHERE TECHNOLOGY

Chapter 1

Role of Microorganisms in Soil Bioremediation

Jean-Marc Bollag, Tawna Mertz, and Lewis Otjen

Laboratory of Soil Biochemistry, Center for Bioremediation and Detoxification, Environmental Resources Research Institute, Pennsylvania State University, University Park, PA 16802

It has long been recognized that microorganisms have distinct and unique roles in the detoxification of polluted soil environments and, in recent years, this process has been termed bioremediation or bioreclamation. The role of microorganisms and their limitations for bioremediation must be better understood so they can be more efficiently utilized. Application of the principles of microbial ecology will improve methodology. The enhancement of microbial degradation as a means of bringing about the *in situ* clean-up of contaminated soils has spurred much research. The rhizosphere, in particular, is an area of increased microbial activity that may enhance transformation and degradation of pollutants. The most common methods to stimulate degradation rates include supplying inorganic nutrients and oxygen, but the addition of degradative microbial inocula or enzymes as well as the use of plants (phytoremediation) should also be considered.

Although "soil bioremediation" is a relatively new term, it describes a phenomenon that has existed since the beginning of life. A wide variety of naturally occurring toxic and recalcitrant organic compounds exist on earth, and many are naturally mineralized The formation of most organic matter begins with plants capable of harnessing the energy of sunlight. This organic matter serves as an important energy source for entire food chains. In a terrestrial ecosystem, the wastes produced by this food chain end up in the soil. Soil organic matter is recycled by a diverse array of soil organisms including bacteria, fungi, actinomycetes, protozoa, earthworms, and insects. Microorganisms are ultimately responsible for mineralizing most organic matter to carbon dioxide. Residual organic matter that is not readily mineralized can be incorporated into humus. By studying these naturally occurring systems of soil bioremediation, researchers should be better able to apply these systems to the clean-up of man-made pollutants.

The use of biological systems to bring about the timely remediation or detoxification of man-made pollutants is the goal of soil bioremediation. The successful implementation of soil bioremediation requires interdisciplinary cooperation among soil biology, soil chemistry, and engineering experts. Although bioremediation is a multidisciplinary undertaking, its central thrust depends on a thorough understanding and utilization of principles of microbial ecology.

Biological treatments of man-made waste have been successfully implemented for decades. Municipal and industrial wastewaters and agricultural wastes are treated in bioreactors as activated sludges, in waste ponds or lagoons, in fixed media reactors using trickling filters, or in fluidized bed reactors containing an active biofilm of microorganisms (*1*) . Bioremediation utilizes the natural role of microorganisms in transformation, mineralization, or complexation by directing those capabilities toward organic and inorganic environmental pollutants.

The United States generates more hazardous waste than any other industrialized nation. Nearly 300 million metric tons annually, or over 1.25 metric tons per person, are produced (original Conservation Foundation estimates). In addition, there are over 1200 Superfund sites, which will incur estimated clean-up costs of $30 billion. Some 40 million citizens who live within four miles of these Superfund sites are at risk. As many as 50,000 additional sites are thought to be contaminated with hazardous compounds and restoration costs for these areas are estimated at $1.7 trillion (*2*). Contaminated environments include surface and subsurface soils and surface and ground waters. Various man-made materials have been dumped at contaminated sites. As a result, many sites contain a complex mixture of contaminants, including petroleum products, organic solvents, metals, acids, bases, brines, and radionuclides. Over a very long period of time, natural degradation activities would eventually destroy some of these contaminants. New affordable technologies are needed to speed up natural remediation processes, thus reducing health risks and restoring natural balances. The repertoire of treatments currently used to remove or destroy contaminants spans physical, chemical, and biological technologies. The use of bioremediation techniques in soils, however, may have some significant advantages over other treatment methods in that cleanup costs may be reduced and site disruption may be kept to a minimum (*3*).

The primary technique that has been used in bioremediation to enhance natural detoxification of contaminated environments is stimulation of the activity of indigenous microorganisms by the addition of nutrients, regulation of redox conditions, and optimization of pH conditions, etc. Other approaches that are still in early stages of testing include (a) inoculation of the sites with microorganisms of specific biotransforming abilities; (b) application of immobilized enzymes; and (c) use of plants to contain or transform pollutants (phytoremediation).

Microbial Activity in Soil

Microorganisms are ubiquitous, inhabiting even the most hostile environments. Their ability to transform virtually all forms of organic material (natural or synthetic) makes them attractive agents of bioremediation. Microbial activity, however, has certain requirements. There are many reasons why organic pollutants may not be quickly degraded when dumped into a soil environment— high concentrations of pollutants, limited amounts of electron acceptors, inadequate supply of nutrients, and unfavorable environmental conditions such as moisture, temperature, pH, ionic strength, or redox status.

In nature, most organic matter is aerobically mineralized, using oxygen as the final electron acceptor. This degradation is usually a stepwise process involving different microorganisms in concert, or in succession, to bring about the mineralization of organic matter. The same is often true for the mineralization of xenobiotics. Slater and Lovatt (4) showed that mixed communities of microorganisms may be more efficient at mineralizing some pollutants, such as chlorinated aromatic hydrocarbons and alkylbenzene sulfonates, than individual species. Sometimes pollutants cannot be directly assimilated by the microbes that oxidize them (co-metabolism), but may instead be further transformed by other populations. These commensal relationships can significantly enhance the mineralization of recalcitrant pollutants and prevent the accumulation of toxic intermediates.

In contrast to the oxygen-dependent metabolism characteristic of aerated portions of soil, other electron acceptors are involved in microbial degradation in anaerobic environments. These reduced environments provide optimal conditions for denitrifying, methanogenic, or sulfate-reducing microorganisms. Although longer incubation times are often required, the ability of anaerobic microorganisms to remove chlorine from environmentally persistent chemicals, such as PCBs (polychlorinated biphenyls), PCE (perchloroethylene), and DDT, may be particularly useful for in situ bioremediation of groundwater. Some compounds, like benzene and related compounds, can be completely mineralized to carbon dioxide or transformed to cell mass under anaerobic conditions (5).

The ability of microorganisms to mineralize pollutants can also be increased through genetic alterations (6). Gene transfer in bacteria occurs naturally through conjugation, transduction, or transformation. Plasmid transfer enables bacteria to adapt quickly to changes in their environment by disseminating the genetic information for the synthesis of the enzymes necessary for degrading new substrates. This characteristic enables bacteria to degrade xenobiotics and is one reason these organisms are useful for controlling the quality of the environment. Experiments have been conducted where in vitro manipulation results in bacterial strains with new metabolic capabilities (7). For instance, pathways have evolved by natural gene tranfer by selection for growth on chloronitrophenols and chlorobiphenyls. Through the use of genetic transfer techniques, it may be possible to introduce the desired plasmids into more suitable hosts for the commercial production of enzymes or inocula.

Soil bioremediation research has, for the most part, focused on the role of microorganisms. It has been shown, however, that plants may also play an important role in the direct and indirect removal of pollutants (*8,9*). Plants can physically remove pollutants from soil by absorbing or translocating them into plant tissue. There, metabolic processes may transform or mineralize pollutants. Plants can also concentrate organic or inorganic pollutants in harvestable portions of the plant; pollutants are then removed by removing the plants. Symbiotic relationships with mycorrhizal fungi may aid in these processes (*10*) . Plants can also indirectly remove pollutants by increasing the biological activity in soil through: (a) rhizosphere interactions; (b) relationships with nitrogen-fixing bacteria; (c) the contribution of dead plant material (leaf litter, etc.); and (d) the provision of suitable habitat for the many other organisms that inhabit the soil.

Humification, the formation of a structureless organic polymer consisting largely of persistent organic substrates, occurs during the microbial decomposition of organic material in soil. In natural soils, humification can be initiated by free radicals, reactive products which are generated by specific microbial enzymes or inorganic catalysts (*11*). The process of humus formation is very complex, because the reactions caused by free radicals occur randomly. It has been found that toxic pollutants can be rendered inert and, thus, be detoxified by their covalent binding to humus (*12*). Therefore, it appears feasible to effectively detoxify pollutants on a large scale by enhancing natural humification processes or by increasing free radical reactions through the use of oxidoreductive enzymes.

It is important to remember that many other organisms (plants, nematodes, earthworms, insects, etc.) exist in soil, but relatively little is known about their contribution to the transformation or mineralization of organic waste (13). More information is needed about their relationships with bacteria and fungi so that complete ecosystems can be restored.

Treatment Technologies

Two types of bioremediation techniques utilize microbes to clean up polluted environments: extraction-treatment techniques and *in situ* treatments. A combined approach using both of these techniques is also possible. Extraction-treatment techniques require removing contaminated soils or groundwater and treating them in a bioreactor or via surface treatments. *In situ* bioremediation techniques generally involve the enhancement of indigenous microbial activity, or inoculation of cultivated microbes into the contaminated environment (*14*).

Extraction-treatment techniques generally require the use of a containment vessel or bioreactor. With a bioreactor, engineers can create optimal conditions for the growth of microorganisms, increase contact between microorganisms and contaminants, utilize specific microbial cultures or enzymes, and manipulate acclimation time for faster biodegradation rates. Bioreactors share the following characteristics: (a) a nutrient delivery system; (b) a blower diffuser system for providing oxygen and mixing; (c) influent and effluent pumps; and (d) the use of water to create an aqueous matrix (*15*). Bioreactors tend to produce gaseous byproducts, such as carbon dioxide, hydrogen, methane, or nitrogen.   These reactors are commonly used to clean up municipal and industrial wastewaters.

In situ bioremediation techniques generally involve enhancing microbial activity in contaminated soil by providing the necessary nutrients, electron acceptors, moisture, etc.   The use of microbial inocula, cell-free enzymes, and plants for in situ bioremediation is still mostly experimental.   To date, most in situ applications have been performed by the petroleum industry to clean up hydrocarbon spills and gasoline tank leaks.   However, in the last few years the possibility of cleaning up regions contaminated with chlorinated hydrocarbons, nitriles, nitrobenzenes, anilines, plasticizers, and others has also been investigated.

Landfarming may be applied if only the upper 0.5 m of the soil profile is contaminated.   In landfarming, biodegradation is enhanced by supplementing the soil with nutrients and oxygen, then tilling and irrigating to create an optimal environment for microbial activity and to increase contaminant/microorganism contact.   Composting is another surface treatment that has been successfully used to remove pollutants.   Composting involves excavating surface soil, mixing the soil with fertilizer and bulking agents, and watering as necessary.   Through landfarming and composting activities, contaminants can often be reduced to acceptable levels.

Sites contaminated with recalcitrant xenobiotics or environments hostile to indigenous microorganisms are potential candidates for inoculation methods. Through the use of classic enrichment techniques or gene transfer technology, large populations of xenobiotic-degrading bacteria may be produced, then inoculated into the contaminated media.   Introducing and dispersing inocula into subsurface soils can be much more difficult than their introduction into surface soils. Microorganisms introduced to the field must compete for resources with native species; they must also elude natural hazards such as microbial toxins and predation.   For this reason, inoculum treatments appear best suited for extreme conditions where competitors and predators are virtually non-existent (16), or for bioreactors where conditions can be optimized for the inoculant (17). *Arthrobacter* sp., *Pseudomonas cepacia*, and *Flavobacterium* sp. have been tried in the field with mixed success.   Repeated inoculations were usually needed to maintain biodegradative activity (13).

In recent years, there has also been much interest in the use of white rot fungi as a soil inoculum (18).   In nature, these fungi are able to decompose wood. Their ability to degrade lignin, one of nature's most persistent organic polymers, led researchers to investigate their ability to mineralize other recalcitrant pollutants.   Indeed, these organisms were capable of degrading a wide variety of persistent pollutants in the laboratory .   Although some success was achieved with field inoculations (19), the problem of predation by and competition with indigenous microflora existed.   It is important to note that these fungi are primarily of the higher orders of basidiomycetes and ascomycetes and have evolved to occupy a rather specific niche as wood-inhabiting and -degrading species.   To expect these organisms to survive, grow, and produce lignin-degrading enzymes in soil without the benefit of their natural environment (wood) is perhaps overly optimistic.

Phytoremediation, the use of plants to detoxify organic or inorganic pollutants, is a promising new area of research (8). Plants may be particularly useful for *in situ* cleanup of soils with shallow contamination (less than two meters deep). The ability of plants to absorb and metabolize pollutants is a well documented phenomenon in the field of weed science. In addition to providing a microbially active rhizosphere, plants can accumulate various pollutants in their vegetative parts, making it possible to harvest the plants and dispose of them through the use of other treatment technologies (9). The ability of plants to accumulate metals is well known, but the use of plants to detoxify contaminated soils is relatively new (6). The influence of the rhizosphere on the degradation of pesticides can be seen when pesticides degrade in agricultural soils. Walton and Anderson (17) documented several indices of microbial activity, showing that trichloroethylene (TCE) in the soil was being degraded by plant root-associated microorganisms. Research on the association between plant roots and microorganisms in the rhizosphere is being actively pursued by scientist interested in applying it to remediation. Plants have been used successfully for wastewater treatment, but their ability to remove xenobiotics has only recently been attributed to the microorganisms associated with their roots .

The use of cell-free enzymes may offer bioremediation specialists an alternative to introducing species of bacteria and fungi into environments that are adverse to microbial growth. The use of microbial enzymes has an advantage over the use of organisms, because enzymes may function under a wider range of conditions than is possible for the growth of organisms. Cell-free enzymes have been used in industry to degrade carbohydrates and proteins and have great potential for *in situ* treatment of environmental contaminants (21). Horseradish peroxidase has also been successfully used to dephenolize coal-conversion wastewaters (22). Problems encountered with the use of enzymes include the difficulty and expense of their extraction and purification, as well as their limited storage life and activity due to the instability of the purified proteins. The stability of enzymes in the environment can be increased by immobilizing them on a solid support. Immobilization of enzymes on fixed-film substrates has been successfully used for wastewater and gas stream treatment. The use of immobilized enzymes offers specificity and rapid reaction times. In the detoxification of wastewaters, water-soluble chemicals can be enzymatically converted to water-insoluble polymers that can be removed via filtration or sedimentation, and the immobilized enzymes may then be subsequently reused.

Aromatic compounds containing hydroxy- or amino-groups are readily bound to soil organic matter (23). Non-specific oxidative reactions catalyzed by enzymes or abiotic agents generate phenoxy radicals or quinonoid structures that are highly reactive and readily form covalent bonds to humic substances. This irreversible binding of chemicals fulfills a beneficial function by detoxifying these compounds. Future studies should concentrate on attempts to enhance detoxification of xenobiotic pollutants in soil by stimulating oxidative binding processes (12).

Advantages and Disadvantages of Bioremediation

A principal reason for the heightened interest in bioremediation is its potential for significantly reducing hazardous waste site cleanup costs. Costs as low as $75 to $200 per cubic yard are reported, compared with conventional technology costs for incineration or secured landfilling of $200 to $800 per cubic yard (3). *In situ* bioremediation offers the added benefit of minimal site disruption, thus reducing public concern. Another important consideration is the relative simplicity of the technology, compared with many other on-site treatment technologies. Operational requirements may be lower than on-site incineration, solidification/stabilization, vitrification, or soil washing systems, possibly resulting in fewer mechanical problems and lower costs.

Despite the potential benefits and relative simplicity of bioremediation systems, a number of pitfalls exist (3,24). The most notable problem areas and issues to consider are: (a) regulatory barriers; (b) scale-up from bench/pilot level to the field; (c) failure of regulatory agencies to consider the full range of remediation options or configurations; (d) liability for failure to achieve goals; and (e) development time and costs. Attempts at bioremediation can also be hampered if the project team does not utilize appropriate and diverse expertise.

It should be noted that regulatory barriers exist at both state and federal levels. The Environmental Protection Agency recognizes, and is taking significant steps to promote, the use of bioremediation through a number of demonstration sites and sponsored programs (25). Nevertheless, projects often run into difficulty when they move from bench-scale to pilot-scale studies, then on to full-scale operation. An understanding of site-specific and chemical-specific limiting factors is critical to successful scale-up. In addition, bioremediation may be very effective in achieving a high percentage reduction of contamination in soil and groundwater, yet still fail to meet stringent clean-up goals that would be more applicable to a technology like incineration. Biological systems tend to slow down as substrate levels diminish, with the microorganisms switching to an alternate energy source. In some cases, microbial degradation may cease when contaminant levels are still above the established goals. At this point, liability becomes an issue (3). Liability disputes may be avoided when site owners and regulators are well versed as to the risks and benefits of using innovative, yet unproven, technology.

Future Considerations

The application of bioremediation technology to decontaminate polluted sites is still a developing science. The mechanisms driving microbial activity and the degradation pathways of specific pollutants need to be further elucidated before successful and better controlled site-specific treatments can occur. Recent advances in biotechnology are capable of modifying organisms at the molecular level for improved degradative performance. This approach has already contributed new tools for analysis and monitoring of complex environmental processes. Other techniques, such as phytoremediation and application of immobilized cells and enzymes, represent novel approaches that may help in treatment of hazardous wastes. A multidisciplinary research approach involving scientists and engineers is needed to provide new strategies for refining available bioremediation techniques. With the cooperation of soil biologists, chemists, and engineers, it should be possible to reduce pollutant concentrations at contaminated sites safely, economically, and efficiently .

Literature Cited

1. Bradford, M. L.; Krishnamoorthy, R., *Chem. Engineer. Progr.* **1991**, 87, 80-85
2. Russell, M.; Colgazier, E. W.; English, M. R.   Hazardous Waste Remediation: The Task Ahead." Waste Management Research and Education Institute.   The University of Tennessee.   R01-2534-19-001-92. 1992.
3. Gabriel, P. F.   *J. Air Waste Manage. Assoc.* **1991**, 41, 1657-1660.
4. Slater, J. H.; Lovatt, D.   In *Degradation of Organic Compounds*; Gibson, D. T., Ed.; Marcel Dekker: New York, 1984.
5. Stroo, H. F.   *J. Environ. Qual.* **1992**, 21, 167-175.
6. van der Meer, J.R., de Vos, W.M., Harayama, S., Zehnder, A.J.B. *Microbiol. Rev.* **1992**, 56, 677-694.
7. Gottschalk, G.; Knackmuss, H.-J., Angew. Chem. Int. Ed. Engl. **1993**, 32, 1398-1408.
8. Cunningham, S. D.; Berti, W. R.   Phytoremediation of Contaminated Soils: Progress and Promise.   Symp. on Bioremediation and Bioprocessing, Div. Petroleum Chemistry, Inc., 205th National Meeting, American Chemical Society, March 18-April 2, 1993.
9. Bell, R. M.   Higher Plant Accumulation of Organic Pollutants from Soils. Project Summary, EPA/600/SR-92/138, U.S.E.P.A., 1992.
10. Donnelly, P. K.; Entry, J. A.; Crawford, D. L.   *Appl. Environ. Microbiol.* **1993**, 59, 2642-2647.
11. Haider, K.; Martin, J. P.; Filip, Z.   In *Soil Biochemistry*, Paul, E.A.and A.D. McLaren, Eds., Marcel Dekker, Inc.   New York, 1975, Vol. 4; pp. 195-244.
12. Bollag, J.-M.   *Environ. Sci. Technol.* **1992**, 26, 1876-1881.
13. Fitter, A.H.   Ecological Interactions in Soil, Plants, Microbes and Animals. Blackwell Scientific Publications; Oxford 1985.
14. Morgan, P; Watkinson, R. J.   *FEMS Microbiol. Rev.* **1989**, 63, 277-300.
15. King, R. B.; Long, G. M.; Sheldon, J. K.   *Practical Environmental Bioremediation.* Lewis Publishers, Boca Raton, 1992.
16. Thomas, J. M., Ward, C. H.   *Environ. Sci. Technol.* **1989**, 23, 760-766.
17. Gibson, D. T.; Sayler, G. S.   Scientific Foundations of Bioremediation - Current Status and Future Needs.   A report from The American Academy of Microbiology.   Washington, DC, 1992.
18. Bumpus, J. A.   In *Soil Biochemistry*; Bollag, J.-M. and G. Stotzky, Eds.; Marcel Dekker, Inc.:   New York, 1993, Vol. 8; pp. 65-100.

19. Lamar, R. T., Dietrich, D. M. *Appl. Environ. Microbiol.* **1992,** 56, 3093-3100.
20. Walton, B. T.; Anderson, T. A. *Cur. Opin. Biotechnol.* **1992,** 3, 267-270.
21. Nannipieri, P.; Bollag, J.-M. *J. Environ. Qual.* **1991,** 20, 510-517.
22. Klibanov, A. M.; Tu, T. M.; Scott, K. P. *Science* **1983,** 221, 259-261.
23. Calderbank, A. *Rev. Environ, Contam. Toxicol.* **1989,** 1108, 71-103.
24. Litchfield, C. D. In Environmental Biotechnology for Waste Treatment; Sayler, G. S.; Fox, R. and J. W. Blackburn, Eds.; Plenum Press: New York, 1991; pp. 147-157.
25. EPA. Bioremediation of Hazardous Wastes. Biosystems Technology Development Program, Office of Research and Development, U.S. Environmental Protection Agency. EPA/600/9-90/041. 1990.

RECEIVED April 19, 1994

Chapter 2

Toxicant Degradation in the Rhizosphere

Barbara T. Walton[1], Elizabeth A. Guthrie[2], and Anne M. Hoylman[3]

[1]Environmental Sciences Division, Oak Ridge National Laboratory, P.O. Box 2008, Oak Ridge, TN 37831–6038
[2]Department of Environmental Sciences and Engineering, School of Public Health, University of North Carolina, Chapel Hill, NC 27599–7400
[3]Graduate Program in Ecology, University of Tennessee, Knoxville, TN 36996–1191

The rhizosphere provides a complex and dynamic microenvironment where bacteria and fungi, in association with roots, form unique communities that have considerable potential for detoxication of hazardous organic compounds. Detoxication may result from degradation, mineralization, or polymerization of the toxicant in the rhizosphere. These detoxication processes are influenced not only by the rhizosphere microbiota, but also by unique properties of the host plant, soil properties, and environmental conditions. Understanding the interactions among plants, rhizosphere microbial communities, and organic toxicants will facilitate the successful use of vegetation to remediate chemically contaminated soils.

A central theme characterizes studies of the degradation of organic toxicants in the plant rhizosphere. This theme is often expressed in the following way: Persistent toxicants in soils constitute a widespread environmental problem, soil microorganisms have a well-recognized ability to detoxicate many organic compounds, and microorganisms abound in the rhizosphere. The presence of plants is known to affect soils and microorganisms in ways that are conducive to natural detoxication (degradation and immobilization) processes. Moreover, existing technologies are inadequate to address the range of environmental problems associated with hazardous waste compounds in soils. These premises lead to the conclusion that vegetation, in conjunction with its associated microbial communities in the root zone, may provide a low-cost, effective strategy for remediation of chemically contaminated soils (*1*).

Despite the ease with which these general tenents of plant-microbe-toxicant studies are formulated, bioremediation in the rhizosphere is a nascent area of investigation in which neither theoretical nor applied aspects are held in general agreement. The potential for vegetation to be used to cleanup soil remains largely a matter of optimistic conjecture. The vast majority of organic toxicants that are candidates for bioremediation have not been studied to determine their degradation rates in the rhizosphere. For those compounds that have been examined in plant-microbe systems, the results may differ from plant to plant, soil to soil, and laboratory to laboratory. Satisfactory explanations for these differences are lacking. Yet the benefits to be realized from using vegetation to support microbial degradation of toxicants in soils are considerable and bioremediation in the rhizosphere warrants critical examination. By investigating plant-microbial associations to better remediate chemically contaminated soils, we may also improve our

0097–6156/94/0563–0011$08.00/0

fundamental understanding of symbiotic relationships in the rhizosphere and the coevolution of plant-microbe-toxicant interactions.

This symposium on "The Rhizosphere and Applications to Bioremediation Technology" is one of the first organized efforts to bring together scientists engaged in research on the degradation of toxicants in the rhizosphere. The syposium, which was held at the 206th Annual Meeting of the American Chemical Society, Chicago IL, August 22-27, 1993, was co-sponsored by the Division of Agrochemicals and the Division of Environmental Chemistry. This overview presents general information about rhizosphere microbiota as it relates to the symposium theme. The influence of plant roots on soil microbiota, the types of bacteria and fungi associated with roots, and the general metabolic degradative capabilities of soil bacteria and fungi are reviewed. Reference is made as appropriate to other chapters of this book for additional information on plant-microbe-toxicant interactions in the rhizosphere.

The Rhizosphere

The "rhizosphere" as described by Hiltner in 1904 is a zone of unique and dynamic interaction between plant roots and soil microorganisms (2). This specialized region is characterized by enhanced microbial biomass and activity. The rhizosphere community consists of microbiota (bacteria, fungi, and algae) and micro- and mesofauna (protozoa, nematodes, mites, and insects) (2). The importance of micro- and mesofauna in decomposition processes in ecosystems (3) suggests that they may also contribute significantly to catabolism of hazardous substances in the rhizosphere, but data on this are scant. Thus, bacteria and fungi in the rhizosphere and their role in catabolism of organic substrates are emphasized in this chapter.

The physical dimensions and microbial activity in the rhizosphere depend on many plant- and site-specific factors. These include the species, age, and vigor of the rooted plant, soil properties, and climatic conditions. The abundance of microorganisms in the rhizosphere is commonly 5 to 20 times, and may be as high as 100 or more times greater than that of non-vegetated soil (4,5) Although bacterial numbers show the greatest augmentation in the rhizosphere, fungal populations may also be increased (2). Site-specific soil parameters, such as temperature, aeration, salinity, texture, and nutrient availability will affect the microbial community in the rhizosphere. Moisture, temperature, and oxygenation are especially critical factors that influence the microbial community and the metabolic pathways that occur in the rhizosphere. For example, anoxic conditions provide favorable microhabitats for denitrification, nitrate ammonification, and methanogenesis (6). The presence of plants may change conditions of soil pH, phosphate and calcium availability, oxygen, CO_2, redox potential, and soil moisture in the rhizosphere (7). The symposium contribution by Sato addresses the role of selected nutrients on pentachlorophenol degradation in soil. These findings may be relevant to toxicant degradation in the rhizosphere, because nutrient concentrations may be higher in the rhizosphere than in nonrhizosphere soil.

Because so many parameters, including the physical dimensions, microbial associations, and soil properties, are affected by the host plant, live plants are an essential component of studies to determine the effect of the rhizosphere on toxicant degradation. Soils removed from the roots cannot be expected to retain rhizosphere properties for extended periods of time.

Plant Roots. Roots provide plants with anchorage and the means to acquire water, nutrients, and other growth substances from soil. In addition, roots provide rhizosphere microorganisms with highly favorable growth conditions. Properties that enable roots to meet the fundamental needs of the plant may also affect the rhizosphere microbial community and the fate of toxicants.

Root morphology contributes to the impact of vegetation on its soil substrate. Features such as root diameter, depth, surface-to-volume ratio, total biomass and surface area vary between and within plant species. Monocotyledons, such as grasses, are more likely to have finely branched terminal roots (less than 100 μm diameter); whereas, coarse terminal roots (between 0.5 to 1.0 mm diameter) are more characteristic of dicotyledons, including most trees. The fibrous root system of monocotyledons is more likely to cover a larger surface area than tap root systems of trees and shrubs. Wheat roots with a mean diameter of 0.1mm, for example, can cover a mean surface area in excess of 6 m^2 (*5*). Plants with fine roots may do well under low nutrient soils, whereas plants with coarse roots may do better in compacted soils (*8*). The roots of perennial species persist much longer than the fine roots characteristic of annual species. Those species that have rapid root turnover may increase the availability of growth substrates to rhizosphere microbiota. Shann and Boyle observe in their symposium contribution that 2,4-D and 2,4,5-T degradation rates were enhanced to a greater extent in soil collected from monocotyledons than from dicotyledons.

In addition to supplying structure for microbial colonization, roots establish and support microbial communities through inputs of photosynthate to the rhizosphere. Sloughed dead cells and plant exudates enrich the habitat of microbiota in the rhizosphere. Organic compounds released from roots include amino acids, vitamins, sugars, tannins, keto acids, and mucigel (*9*). Other secretions are mucilages, enzymes (e.g., peroxidases and lysases), growth regulators, CO_2, and ethylene (*10,11*). These root exudations are an important cause of the greater microbial biomass present in the rhizosphere compared to nonrhizosphere soil (*12.*).

The ability of plants to release photosynthate to soil through exudation and sloughing of dead cells increases soil organic matter. This greater organic content of the rhizosphere soil may alter toxicant sorption, bioavailability, and leachability. For example, rhizosphere microorganisms may facilitate copolymerization of toxicants with humic substances through biogenic polymerization reactions that occur in the formation of humics (*13,14*). Data presented in this symposium by Dec and Bollag on chlorophenols and on PAHs by Walton and coauthers indicate that associations of both groups of compounds with soil organic fractions may be influenced either directly or indirectly by microbiota in the rhizosphere.

The effects of plants and plant roots on bioremediation processes are specifically addressed in recent reviews (*1, 15, 16*) and in symposium presentations by Shann and Boyle; Qui, et al.; Knaebel and Vestal; Watkins, et al.; Davis, et al.; and Anderson, et al.

Plant Microbial Interactions. The species of plant host is a critical factor in the development of a rhizosphere community (Table I). Bacteria in the rhizosphere are typically gram negative, non-sporulating bacilli plus a low proportion of gram positive bacilli, cocci, pleomorphic rod forms and aerobic non-spore forming bacteria (Table II). Motile bacteria capable of rapid growth, such as *Pseudomonas*, are common in rhizosphere soils. Pseudomonads appear to be well adapted to high root exudate levels and readily utilize organic substances released by the host plant. Other species, such as *Arthrobacter*, are abundant at low root exudation levels. *Agrobacterium, Achromobacter,* ammonifying, denitrifying, sugar-fermenting, and cellulose-decomposing bacteria are also present in the rhizosphere in higher numbers than in nonrhizosphere soils (*22, 23, 24*).

Perhaps, the best characterized bacterial associations with plant roots are those of nitrogen-fixing bacteria (rhizobia) and leguminous plants (e.g., peas, soybean, alfalfa). These symbiotic associations are unique for both their nutrient contribution to soils and the biochemical/physiological relationship between bacteria and host plant. The bacteria invade the root tissues, which respond by formation of nodules that become the site where bacteria fix atmospheric nitrogen. The formation of nitrogen-fixing nodules

Table I. Associations Between Nitrogen-fixing Prokaryotes and Vascular Plants

Prokaryote		Vascular Plant	
Class	Genus	Class/Order	Family/Genus/Common name
Cyanobacteriales	*Anabaena*	Pteridophyta	*Azolla*, aquatic fern
	Nostoc	Cycadales	All genera examined
	Nostoc	Angiospermae	*Gunneral* spp., creeping plants and shrubs
Actinomycetales	*Frankia*	Angiospermae	Various non-leguminous genera, green alder, silverberry shrub, sweet fern, raspberry shrub
Eubacteriales	*Rhizobium*	Angiospermae	Ulmaceae, *Parasponia*, elm, tropical plants
			Leguminosae, peas, kidney bean, clover, soya bean, cowpea, lentil, lucerne
Eubacteriales	*Azospirillum*	Angiospermae	cereals, grasses, tomato

Compiled from ref. *17, 18, 19, 20.*

Table II. Common Soil Bacteria

Family/Genus	Description	Location/Function
Azospirillum	spiral or curved bacteria, Gram negative	rhizosphere; nitrogen fixers in tropical grasses
Pseudomondaceae	flagellated rods, aerobic or facultative anaerobes	decompose soluble compounds from organic matter
Azotobacteraceae	obligate anaerobes	free-living N_2 fixers
Rhizobiaceae		
Rhizobium	root-infecting bacteria	produce nodules on leguminous plants; fix N_2
Agrobacterium		produce galls; fixes N_2
Enterobacter (Aerobacter)	Gram-negative, facultatively anaerobic rods	
Nitrobacteraceae	heterogeneous, motile and non-motile	oxidize reduced forms of inorganic nitrogen
Endospore-forming bacteria		
Bacillus	aerobic or facultative anaerobic	many species fix N_2; produce extracellular enzymes that hydrolyse complex organic molecules
Clostridium	anaerobic	ferment cellulose, starch, pectin, sugars
Actinomycetes	Gram-positive, irregular morphology	largest and heterogeneous group
Arthrobacter	pleomorphic	resistant to dessication and starvation
Nocardia, Actinomyces	mycelial, saprophytes, filamentous	
Streptomycetes	aerobes	utilize a variety of organic compounds produce antibiotics
Frankia		fix N_2 in non-leguminous angiosperms

Compiled from ref. *21, 3.*

results from chemotactic communication between plants and microbial populations. Release of chemoattractants such as flavenoid compounds attract rhizobia to the root surface where they bind to susceptible root hairs with the aid of lectins (plant proteins). Tryptophan is released, then oxidized by rhizobia to initiate root hair curling for nodule formation (5). Although nitrogen-fixing bacteria are not the most numerous members of the rhizosphere microbial community, their role in nitrogen availability to the plant is clearly important (25).

Fungi are the second most prevalent group of microorganisms in soil and, occasionally, fungal biomass exceeds bacterial biomass. Plant benefits from these fungal associations include enhanced water and mineral uptake, increased resistance to pathogens, and tolerance to many environmental stresses (26). Non-mutualistic fungi (e.g., *Fusarium spp.*) also occur in the rhizosphere and rhizoplane (5).

There are three major types of soil fungi: (1) pathogenic root-infecting fungi, (2) saprophytic, and (3) mycorrhizal fungi. Mycorrhizal fungi form mutualist associations that integrate plant roots with fungal mycelia to form unified operational cell entities (Table III). Mycorrhizal fungi that enter root cells (endomycorrhizae) are distinquished from mycorrhizal fungi that only surround roots and extend into intercellular spaces (ectomycorrhizae) (12). Mycorrhizal fungi are especially important because they are associated with the rhizospheres of a wide range of herbaceous and woody plants (Table III) and may improve plant success under nutrient- or water-limited conditions (21). In addition, Mycorrhizal communities may protect plants from protentially toxic substances in soils via the physical barrier created to the root or by means of fungal secretions (organic acids, antibiotics) and detoxication enzymes. Donnelly and Fletcher provide an excellent review of mycorrhizal fungi, their degradative capabilities, and their potential use in bioremediation.

Rhizosphere Microbiota and Toxicant Interactions

Plants are exposed to naturally occurring toxicants, including phenols, terpenes, and nitrogen-containing alkaloid compounds (32) and potentially toxic anthropogenic compounds in soils. Plants may be protected by pectins and lignins that may exclude or sorb high molecular weight, lipophilic compounds (32,33), thus preventing their entry into the root. In addition, plants secrete organic compounds (e.g., phenols, phytoalexins) that are bacteriocidal, fungistatic, or inhibitory (alleopathic) to other species. Microbiota in the rhizosphere are able to degrade many of these naturally occurring allelopathic compounds and anthropogenic organic substances (1). In another chapter of this book, Walton and coauthors propose, that plants may be protected against potentially toxic compounds in soils by the metabolic detoxication capabilities of the rhizosphere microbial community. They propose that the exchange of nutritional benefits from the host plant to the rhizosphere microbial community for the rapid, versatile detoxication capability of the microbial community may constitute the basis of a mutually beneficial, dynamic relationship between the host plant and its rhizosphere microbial community.

The rhizosphere provides microhabitats for aerobic and anaerobic microorganisms. Although most plants grow under the aerobic conditions of unsaturated soils, which permit roots to utilize oxygen in the soil for respiration, plants can also withstand transient anaerobic conditions that may occur during periods of high rainfall or flooding. Anoxic states may exist for more prolonged periods in anaerobic centers of aggregated soil particles or under wetland conditions, where wetland plant species have physiological mechanisms to meet the oxygen demands of the roots.

The oxygen status of the rhizosphere is likely to be an important determinant of the microbial composition of the rhizosphere. Because both aerobic and anaerobic microorganisms can be expected to be present in the rhizosphere, microbial degradation of toxicants by both catabolic pathways is relevant (Table IV). In their symposium

Table III. Associations Between Mycorrhizae and Vascular Plants

Fungus	Vascular Plant		
Family/Genus	Order	Family	General Description
Ectomycorrhizae			
Ascomycetes	Angiospermae	Dipterocarpaceae	tropical timber trees
Fungii Imperfectii		Fagaceae	beeches, oaks, chestnuts trees
Zygomycetes		Myricaceae	myrtle and eucalyptus tree
Basidiomycetes		Salicaceae	willow trees
		Betulaceae	alders, birches, ironwood trees
		Leguminosae	clover, alfalfa, locust and acacia trees, cow peas
	Gymnospermae	Pinaceae	spruce, pines, hemlocks trees
Endomycorrhizae			
Vesicular-arbuscular (VA)			
Endogonaceae	Angiospermae	Leguminosae	beans
		Rosaceae	apple tree
		Solanaceae	potatoe
	Gymnospermae	Pinaceae	pines, spruce tree
Non-vesicular			
	Ericales	Ericaceae	heath, arbutus, azalea, rhododendron, laurel
		Empetraceae	rosemary, crowberry shrubs

Compiled from ref. *19, 27, 28, 29, 30, 31.*

Table IV. Substrates Degraded by Bacteria in the Rhizosphere

Genus	Substrate	
	Aerobic degradation	Anaerobic degradation
Achromobacter	hydrocarbons	
Acinetobacter	hydrocarbons	
Agrobacterium-Rhizobium	hydrocarbons; halogenated aromatics	
Alcaligenes	aromatic hydrocarbons; halogenated aromatics; hydrocarbons	halogenated aromatics
Azotobacter	aromatic hydrocarbons; halogenated aromatics	
Flavobacterium	hydrocarbons	
Methanobacterium		halogenated aliphatics
Mycobacterium	halogenated aliphatics	
Nitrosomonas	halogenated aliphatics	
Nocardia	aromatic hydrocarbons; halogenated aromatics; hydrocarbons	
Pseudomonas	aromatic hydrocarbons; halogenated aromatics; hydrocarbons	halogenated aromatics
Rhodopseudomonas		halogenated aromatics
Xanthobacter	halogenated aliphatics	

Compiled from ref. 27, 34, 35, 36.

presentation, Barkovskii and coauthors provide evidence that polyphenolic substances are transformed by *Azospirillum lipoferum*, found in the rhizosphere of rice, and propose that polyphenolics may be alternative electron acceptors.

Biologically degradable waste chemicals released during the production and use of fossil fuels as well as many organic solvents, including chlorinated aliphatics, are excellent candidates for bioremediation in the rhizosphere. Some agrochemicals may also degrade in the rhizosphere; however, herbicides may have a low potential for bioremediation in the rhizosphere because these compounds were chosen for commercial development specifically because of their toxic properties to plant life. The efficacy of these herbicides may derive in part from their recalcitrance to degradation in the rhizosphere. That is, the inability of a microbial community to effectively detoxicate a herbicide in the rhizosphere, in some instances, may contribute to the selection of the compound for commercial development. Conversely, benzene, toluene, xylenes, chlorinated solvents and some polycyclic aromatic hydrocarbons (PAHs) are good candidates for bioremediation in the rhizosphere. Gibson (*37*) provides an excellent introduction to microbial degradation of organic compounds in soil and water and, in so doing, provides a framework for considering the range of degradation pathways that may occur in the rhizosphere. Some observations on fungal and bacterial degradation of organic compounds are presented below.

Fungi possess a wide range of enzymes that decompose plant material, including lignin and cellulose. Examples include cellulase, lignase, hydrolytic acid protease, peroxidase, lipase, polyphenol oxidase, and cytochrome P-450 monooxygenase (*38*). Fungal enzymes have been shown to degrade numerous anthropogenic pollutants such as polycyclic aromatic hydrocarbons (PAHs), phenols, polychlorinated biphenyls, nitroaromatics, cyanide, chlorinated dibenzodioxins, herbicides, and chlorinated pesticides (*35, 38, 39, 40*). *Cunninghamella elegans, Rhizoctonia solani*, and *Phanerochaete chrysosporium* metabolize PAHs (Fig. 1) via multiple enzymes. Although there are fundamental differences between degradation of aromatic compounds by eukaryotes and prokaryotes, oxidation of the PAH phenanthrene by *P. chrysosporium* yields metabolites similar to phenanthrene metabolites produced by bacteria (*35*), indicating that several enzymatic pathways may exist for fungal metabolism of PAHs.

Prokaryotes typically degrade PAHs and other aromatic hydrocarbons aerobically by incorporating two oxygen atoms into the ring structure via dioxygenase, whereas, eukaryotes more commonly incorporate a single oxygen atom via monooxygenases (*34*). The difference between these two catabolic pathways are most evident at the dihydrodiol step in metabolism (Fig. 1). Bacteria oxidize aromatic compounds to a *cis*-dihydrodiol (or catechol), whereas eukaryotes (e.g., fungi) hydroxylate the aromatic ring to a *trans*-dihydrodiol. Further bacterial metabolism of the aromatic ring requires the presence of two hydroxyl groups in the *ortho* or *para* positions (*35*). Biodegradation of PAHs is the primary route by which di- and tri-cyclic PAHs are removed from soil systems. However, in general, higher molecular weight PAHs are resistant to microbial degradation in the absence of alternative carbon sources (*35*). In a symposium presentation by Watkins and coauthors, mineralization of naphthalene was shown to decrease in vegetated microcosms compared with microcosms without vegetation. Watkins and coauthors also found that naphthalene volatilization was promoted by grass. In studies using alfalfa and toluene, however, Davis and coauthors found the presence of plants promoted biodegration and reduced volatilization of naphthalene.

Halogenated compounds undergo both aerobic and anaerobic degradation by bacteria and fungi of several different genera (*41*). Recent publications provide comprehensive overviews of degradative pathways under many environmental conditions (*41, 42, 43, 44, 45*). In general, microbial metabolism of halogenated compounds, such as chlorinated insecticides and halogenated alkanes and alkenes, is highly dependent upon both the quantity and position of the halogen substituent(s), as well as the presence of specific microbial growth substrates (*43, 46, 47*). Degradation of halogenated organic

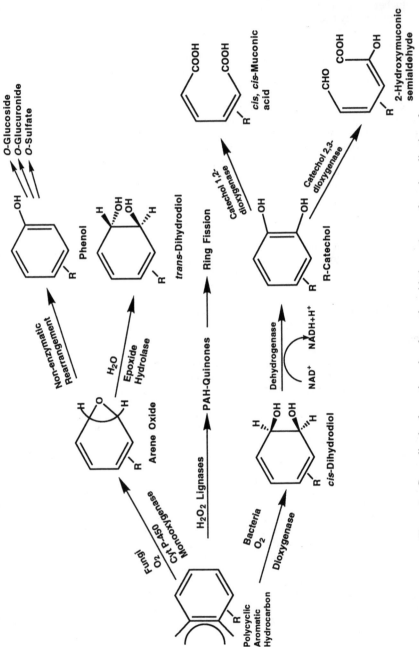

Figure 1. Generalized eukaryotic and prokaryotic oxidative pathways for mineralization of polycyclic aromatic hydrocarbons (adapted from 34, 35).

compounds occurs in pure cultures or consortia of microorganisms that utilize the compound for energy, as well as in cultures of microorganisms that cometabolize the substance through nonspecific enzymes (e.g., methane monooxygenase, toluene dioxygenase) *(43, 48, 49).* Cometabolism occurs without either incorporation of carbon into microbial biomass or energy utilization from the substrate; the compound is degraded but not used for growth or energy *(5, 50).* Degradation of trichloroethylene (TCE) has been observed in microorganisms from diverse aquatic and soil environments, including rhizosphere soils *(51).* Aerobic microbial degradation of TCE has been demonstrated in cultures of methanotrophic organisms that utilize methane for energy and in microbial cultures that grow on aromatic compounds such as phenol or toluene *(52, 53).*

A diverse group of bacteria participate in the anaerobic degradation of organic compounds (Table 1) and many of these may also be present in the rhizosphere. Nitrate- and sulfate-reducing bacteria, acetagens, and methanogenic consortia are examples of bacteria capable of anaerobic degradation of various hazardous chemicals *(54).* Nitrate-reducing microorganisms have been isolated that degrade substituted aromatic compounds such as benzoate (Fig. 2). Other microorganisms have been isolated that carry out reductive dehalogenation, the only known biodegradation mechanism for some highly chlorinated compounds such as tetrachloroethylene (PCE) (Fig. 3) *(44).* However, degradation of PCE under anaerobic conditions can result in the accumulation of vinyl chloride, which is of greater toxicological concern than PCE. Because vinyl chloride is mineralized aerobically, the complete mineralization of PCE may be accomplished by the sequential actions of anaerobic and aerobic microorganisms Whether such conditions are met in the rhizosphere as a result of the sequential wetting and drying that occurs as a result of rain events followed by evaporation, runoff, and water usage by plants has not been examined, but provides another interesting possibility for bioremediation in the rhizosphere.

Summary

Studies of the interactions between plant roots and their associated microbiota have provided the basis for many insights in the areas of plant pathology and agronomy. These insights, in turn, led to improvements in preventing plant disease and increasing plant productivity. Furthermore, the unraveling of bacterial and fungal associations with plant roots has enhanced understanding in fundamental areas of biology and ecology, including symbiosis, gene transfer, coevolution, nitrogen fixation, energy flow, and nutrient cycling . In recent years, the possibility that plant roots may enhance microbial degradation of organic chemicals in soils provides a new approach to solving environmental problems related to waste chemicals in soils any may also provide an avenue for further understanding of fundamental biological processes.

The use of vegetation to accelerate mineralization of waste chemicals in soils offers the advantages of a solar-powered treatment system that is aesthetically pleasing and can end legal liability for chemical wastes when mineralization is achieved. Thus, the use of vegetation for remediation of soils offers distinct advantages over existing approaches to remediation technologies that merely transfer the waste, as well as legal liability for that waste, from one site to another. The use of vegetation for remediation and reclamation has the added attraction of a higher potential for public acceptance than many existing technologies, such as incineration, excavation, and long-term storage.

The papers presented at this symposium on "The Rhizosphere and Applications to Bioremediation Technology" provide impressive testimony to the variety of experimental approaches, possible limitations, and intriguing possibilities for bioremediation in the rhizosphere.

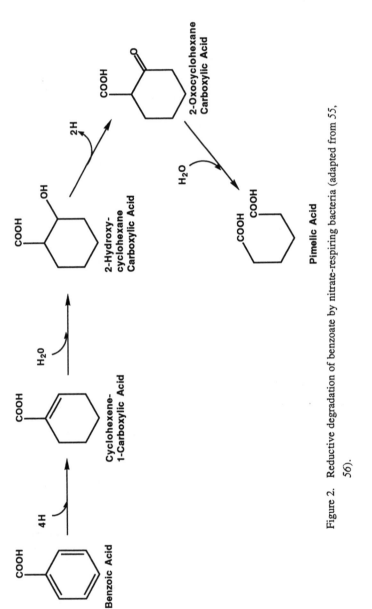

Figure 2. Reductive degradation of benzoate by nitrate-respiring bacteria (adapted from 55, 56).

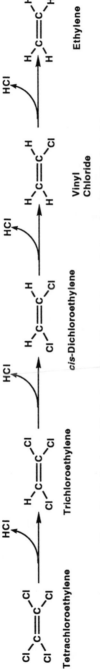

Figure 3. Reductive dechlorination of tetrachloroethylene to ethylene as mediated by bacterial transition-metal coenzyme F430 and vitamin B12 (adapted from 39).

Acknowledgments

We are grateful to F. K. Pfaender, University of North Carolina, Chapel Hill and A. V. Palumbo and N. T. Edwards, Environmental Sciences Division, Oak Ridge National Laboratory, Oak Ridge, TN for helpful comments on this manuscript. Oak Ridge National Laboratory is managed by Martin Marietta Energy Systems, Inc., for the U. S. Department of Energy under Contract No. DE-AC05-84OR21400. Publication No. 4271 of the Environmental Sciences Division, ORNL.

Literature Cited

1. Anderson, T.A.; Guthrie, E.A.; Walton, B.T. *Environ.Sci. Technol.* **1993**, *27*, 2630-2635.
2. Curl, E.A.; Truelove, B. *The Rhizosphere.* Springer-Verlag. Berlin, **1986**, 288p.
3. Swift, M.J.; Heal, O.W.; Anderson, J.M.; *Decomposition in Terrestrial Ecosystems*; University of California Press: Berkeley and Los Angeles, CA, **1979**.
4. Katznelson, H. *Soil Sci.* **1946**, *62*, 343-354.
5. Atlas, R.; Bartha, R. *Microbial Ecology: Fundamentals and Applications*; The Benjamin/Cummings Publishing Company, Inc: Menlo Park, CA. **1993**.
6. Ehrlich, H. In *Soil Biochemistry*; Bollag, J.M.; Stotsky, G., Eds.; Marcel Dekker, Inc.: Dekker, NY, **1993**, Vol. 8, pp. 219-234.
7. Foster, R.C.; Rovira, A.D.; Cook, T.W. *Ultrastructure of the Root-Soil Interface.* American Phytopathological Society: St. Paul, MN, **1983**.
8. Fitter, A.H. In *Plant Roots: The Hidden Half*; Waisel,Y.; Eshel, A; Kafkafi, U., Eds.; Marcel Dekker, Inc.: New York, NY. **1991**, pp. 3-24.
9. Lynch, J.M.; Whipps, J.M. In *The Rhizosphere and Plant Growth;* Keister, D. L.; Cregan, P.B., Eds.; Kluwer Academic Publishers: Boston, MA. **1991**, pp. 15-24.
10. Bowen, G.D.; Rovira, A.D. In *Plant Roots: The Hidden Half*; Waisel,Y.; Eshel, A; Kafkafi, U., Eds.; Marcel Dekker, Inc.: New York, NY. **1991**, pp. 641-670.
11. Richards, B.N. *The Microbiology of Terrestial Ecosystems*; John Wiley & Sons: New York, NY. **1987**.
12. Vogt, K.A.; Bloomfield, J. In *Plant Roots: The Hidden Half*; Waisel,Y.; Eshel, A; Kafkafi, U., Eds.; Marcel Dekker, Inc.: New York, NY. **1991**, pp. 287-308.
13. Christman, R.F.; Oglesby, R.T. In *Lignins*, Sarkanen, K.V.; Ludwig, C.H., Eds.; Wiley-Interscience: New York, NY. **1971**, pp. 769-795.
14. Hedges, J.I.. In *Humic Substances and Their Role in the Environment* ; Frimmel, F.H.; Christman, R.F., Eds.; John Wiley & Sons: New York, NY. **1988**, pp. 45-58.
15. Shimp, J.F.; Tracy, J.C.; Davis, L.C.; Lee, E.; Huang, W.; Erickson, L.E.; Schnoor, J.L. *Crit. Rev. Environ. Cont.* **1993**, *23*,41-77.
16. Davis, L.C.; Erickson, L.E.; Lee, E.; Shimp, J.F.; Tracy, J.C. *Environ. Prog.* **1993**, *12*, 67-75.
17. Commandeur, L.; Parsons, J.R. *Biodegrad.* **1990**, *3*, 1207-1220.
18. Allen, M.F. *The Ecology of Mycorrhizae*; Cambridge University Press: New York, NY, **1991**.
19. Sprent, J. I. In *Advanced Plant Physiology*, Wilkins, M.B., Ed.; John Wiley & Sons: New York, NY. **1987**, pp. 249-280.
20. Sprent, J. *Nitrogen Fixing Organisms - Pure and Applied Aspects;* Chapman and Hall: New York, NY, **1990**.
21. Mukerji, K.G.; Jagpal, R.; Bali, M.; Rani, R. In: *Plant Roots and Their Environment - Proceedings of an ISRR-Symposium;* McMichael, B.L.; Persson, H., Eds.; Elsevier Science Publisher: New York, NY, **1991**, pp. 299-308.

22. Campbell, R.; *Plant Microbiology*, Edward Arnold: Baltimore, MD, **1985**.
23. Vancura, V.; Kunc, F.; *Interrelationships Between Microorganisms and Plants in Soil*; Elsevier: Amsterdam, **1989**.
24. Nakas, J.P.; Hagedorn C.; *Biotechnology of Plant-Microbe Interactions*; McGraw-Hill Publishing Company: New York, NY, **1990**.
25. Campbell, R.M.; Greaves, P. In *The Rhizosphere;* Lynch, J.M.,Ed.; John Wiley and Sons: New York, NY. **1990**, pp 11-34.
26. Harley, J.L.; Smith, S.E. *Mycorrhizal Symbiosis*; Academic Press: New York, NY, **1983**.
27. Radford, A. E.; Ahles, H. E.; Bell, C.R. *Manual of the Vascular Flora of the Carolinas*; The University of North Carolina at Chapel Hill.: Chapel Hill, NC, **1981**.
28. Postgate, J.R. *The Fundamentals of Nitrogen Fixation*; Cambridge University Press: New York, NY, **1982**.
29. Reid, C.P.P. In *The Rhizosphere;* Lynch, J.M., Ed.; John Wiley and Sons: New York, NY, **1990**, pp 281-316.
30. Hutchinsen, J. *The Families of Flowering Plants*; Oxford University Press: London, **1973**. 968p.
31. Smith, J.P. *Vascular Plant Families*; Mad River Press: Eureca, CA, **1977**.
32. Taiz, L.; Zeiger, E. *Plant Physiology*. The Benjamin/Cummings Publishing Co. Inc.; Redwood City, CA, **1991**.
33. Harms, H.H. *Pestic. Sci.* **1992**, *35*, 277-281.
34. Varanasi, I.U. *Metabolism of Polycylic Aromatic Hydrocarbons in the Aquatic Enviroment;* CRC Press, Inc.: Boca Raton, FL. **1989**.
35. Cerniglia, C.E. *Biodegrad.* **1992**, *3*, 351-368.
36. Gantzer, C.J.; Wackett, L.P. *Environ.Sci. Technol.* **1991**, *24*, 715-722.
37. Gibson, D.T. *Microbial Degradation of Organic Compounds*; Marcel Dekker, Inc.: New York, NY, **1984**, 535 p.
38. Wainwright, M. In *The Fungal Community: Its Organization and Role in the Ecosytem;* Carroll, G.; Wicklow, D., Eds.; Marcel Dekker, Inc.: New York, NY. **1991**.
39. Aust, S.D., *Microb. Ecol.* **1990**, *20*, 197-209.
40. Barr, D. P.; Aust, S. D. *Environ.Sci. Technol.* **1994**, *28*, 78A-87A.
41. Chaudhry, G.R.; Chapalamadugu, S. *Microbiol. Rev.* **1991**, *55*, 59-79.
42. Belkin, S. *Biodegrad.* 1992, 3, 299-313.
43. Pfaender, F.K. In *Significance and Treatment of Volatile Organic Compounds in Water Supplies*; Ram, N.M.; Christman, R.F.; Cantor, K.P., Eds.; Lewis Publishers, Inc.: Chelsea, MI. **1990**, pp. 205-226.
44. Mohn, W.W.; Tiedje, J.M. *Microbiol. Rev.* **1992**, *56*, 482-507.
45. Murray, W.D.; Richardson, M. Crit. Rev. Environ. Sci. Technol. 1993, 23, 195-217.
46. Kuhn, E.P.; Suflita, J.M. In *Reactions and Movement of Organic Chemicals in Soils*; Sawhney, B.L.; Brown, K., Eds.; Soil Science Society of America and American Society of Agronomy: Madison, WI. **1989**, pp. 111-180.
47. Fan, S; Scow, K.M. *Appl. Environ. Microbiol.* **1993**, *59*, 1911-1918.
48. Janssen, D.; Witholt, B. In *Metal Ions in Biological Systems, Degradation of Environmental Pollutants by Microorgansims and Their Metalloenzymes;* Sigel, H.; Sigel, A., Eds.; Marcell Dekker, Inc.: Dekker, NY, **1992**, Vol. *28*, pp. 299-327.
49. Grbic-Galic, D. 1990. In: *Soil Biochemistry*; Bollag, J.M.; Stotsky, G., Eds.; Marcell Dekker, Inc.: Dekker, NY, **1993**, Vol. *6*, pp. 117-189.
50. Dalton, H.; Stirling, D.I. *Phil. Trans. R. Soc. Lond. B*, **1982**, *297*, 481-496.
51. Walton, B.T.; Anderson, T.A. *Appl. Environ. Microbiol.* **1990**, *56*, 1012-1016.
52. Wackett L.P.; Gibson, D.T. *Appl. Environ. Microbiol.* **1988**, *54*,1703-1708.

53. Nelson, M.J.K.; Montgomery, S.O.; Mahaffey, W.R.; Pritchard, P.H. *Appl. Environ. Microbiol.* **1987**, *53*, 949-954.
54. Freedman, D.; Gossett, J. *Appl. Environ. Microbiol.* **1991**, *57*, 2847-2857.
55. Young, L. In *Microbial Degradation of Organic Compounds*; Gibson, D. T., Ed.; Plenum Press: New York, NY, **1984**, pp. 487-523.
56. Rochkind, M.L.; Blackburn, J.W.; Sayler, G.S. *Microbial Decomposition of Chlorinated Aromatic Compounds*; Hazardous Waste Engineering Research Laboratory, U.S. Environmental Protection Agency, Cincinnati, OH. EPA/600/2-86/090, **1986**, 269p.

RECEIVED May 19, 1994

Interactions among Microorganisms, Plants, and Chemicals

Chapter 3

Polyphenolic Compounds Respired by Bacteria

Andrei L. Barkovskii[1], Marie-Louise Boullant,
and Jacques Balandreau

Laboratoire d'Ecologie Microbienne du Sol, Unité de Recherche Associée,
1450 Centre National de la Recherche Scientifique, Université Claude
Bernard-Lyon 1, Boulevard du 11 Novembre 1918, 69622 Villeurbanne
Cedex, France

The effects of 6 phenolics were studied on 2 rice rhizosphere strains of
Azospirillum lipoferum. No effect was observed with strain 4B.
Under 5 kPa O_2, growth of strain 4T was stimulated by phenolics,
especially caffeic (CAF) acid, from which lipophilic compounds were
formed. Under nitrogen-fixing (ARA) and electron acceptor-limiting
conditions, strain 4T could reduce acetylene only in the presence of
both CAF and a reoxidizing agent (N_2O); cell-free extracts showed a
simultaneous hypsochrome shift in the UV spectrum. It is suggested
that strain 4T possesses the unique ability to use polyphenolics as
alternative terminal respiratory electron acceptors under low oxygen.
Such respiratory transformations of phenolics could significantly
contribute to the fate of xenobiotics in the environment.

Rapid degradation of xenobiotics in the rhizosphere of plants (*1-4*) provides a possible
opportunity to apply these processes to the use of specific plant-bacteria systems for
soil remediation. Symbiotic and rhizospheric bacteria are good candidates for this
purpose.

Phenolic compounds are part of root exudates and intermediary products in the
metabolism of molecules containing aromatic rings (lignin, tannins, many pesticides)
(*5*). Phenolic compounds can enter the soil environment from other sources, such as
the generation of wastes by petroleum-related (*6*) and aniline-related (*7*) industries.

The genera *Rhizobium* and *Bradyrhizobium* are probably the most studied
symbiotic N_2 fixers. The ability of rhizobia to degrade aromatic compounds is well-
known (*8-10*), and their ability to degrade some haloaromatic compounds has also
been demonstrated (*11*).

Azospirillum spp. is probably the most studied non-symbiotic N_2 fixer because
of its well-known property of root-growth promotion demonstrated on a very broad
range of plant species. These bacteria efficiently colonize the elongation and root-hair
zones of wild plants as well as cereals, tomato, pepper, cotton, soybean, and
sunflower plants (*12-14*). *Azospirillum* could present numerous advantages for a
successfull introduction for soils decontamination. In particular, they already fulfill

---

[1]Current address: Department of Civil and Environmental Engineering, University
of Michigan, 219 Environmental and Water Resource Engineering Building,
Ann Arbor, MI 48109–2125

0097–6156/94/0563–0028$08.00/0

industrial, commercial, and legal requirements for their utilization in agriculture (*15*). In our previous work (*11*), we studied the catabolism of phenolics by the genus *Azospirillum*. It was observed that about 40% of tested *Azospirillum* strains were able to degrade benzoate or phenol in the presence of a co-substrate under a wide range of aeration conditions. In contrast, most studies on the aerobic degradation of phenolics by bacteria (*Pseudomonas, Alcaligenes, Acinetobacter*) demonstrated that phenolic consumption is strongly dependent on aeration and can occur in the absence of a co-substrate (*16-19*).

Azospirillum strains need special oxygen conditions to express nitrogenase activity. It has been demonstrated (*20*) that maximum nitrogen fixation (ARA) occurs under O_2 concentrations close to 1%, and that *Azospirillum* nitrogenase was inhibited when O_2 concentration reached 8%. Such conditions may be encountered in the rhizosphere of many crops (*21, 22*) where nitrogen-fixing microorganisms find both a suitable supply of oxidizable compounds (e.g. carbohydrates) (*23*) and the terminal respiratory electron acceptors required to efficiently produce the energy necessary to sustain growth and ATP demanding nitrogen-fixing activity (*24*). When O_2 becomes limiting, other terminal respiratory electron acceptors such as NO_3^- are often used.

The easy oxidation-reduction changes of polyphenolic compounds have been shown by many studies to be involved in biochemical reactions, especially with quinone coenzymes (*25, 26*). Soil phenolic compounds are also chemically involved in electron transfer to metals (*27, 28*) or to organic compounds (*29, 30*) and are transformed during those reactions. In every case, they exhibit typical reversible reactions linked to electron transfers (*31*).

All these facts prompted us to undertake a study of the comparative effects of a series of simple phenolic compounds on the metabolism of motile (4B) and non-motile (4T) *Azospirillum lipoferum* strains. The interaction between *Azospirillum lipoferum* strains and phenolic compounds was investigated under various levels of O_2 to elucidate the possible involvement of these compounds in electron transfers in the root environment and their possible transformations by bacterial respiration. An advantage of using a nitrogen-fixing aerobic bacterium is that nitrogenase activity requires a high supply of energy and thus relies mostly on ATP generated through the respiratory electron transfer chain. Moreover, it can be measured (through ARA, Acetylene Reduction Activity) with tremendous sensitivity, making acetylene reduction a more sensitive method than growth measurements to monitor metabolic activities of N_2-fixing bacteria. If metabolic activities of the bacteria are limited by the concentration of a terminal electron acceptor, ARA depends only upon its concentration.

In the experiments reported here, possible electron acceptors are O_2 or nitrous oxide (N_2O). Nevertheless, under our conditions, N_2O is unable to accept electrons from bacterial N_2O reductase because this enzyme is inhibited by acetylene concentrations well below the 10% acetylene used for nitrogenase activity measurements (*32*).

In the course of this study, we also tested the hypothesis that a combination of a phenolic compound with a reoxidizing agent such as N_2O would generate oxidized compounds usable as terminal respiratory electron acceptors. The presence of acetylene prevented the transfer of electrons to N_2O.

Material and Methods

Phenolic Molecules. Six phenolic compounds with increasing sidechain length were assayed: Catechol (CAT); 4-Hydroxybenzoic acid (4-HBA); 3,4-Dihydroxybenzoic acid (3, 4-DHBA); 3-Methoxy-4-Hydroxybenzoic acid (Vanillic acid, VAN); 2-Hydroxyphenylacetic acid (2-HPAA), and 3,4-Dihydroxycinnamic acid (Caffeic acid, CAF). CAT, 3,4-DHBA and CAF rings carry ortho hydroxyl groups, whereas 4-HBA, 2-HPAA and VAN rings carry only one hydroxyl group in the ortho or para position relative to the acidic side chain.

Bacterial Strains. *Azospirillum lipoferum* 4B and 4T strains were isolated from the rhizosphere of rice (*33*). The two strains differ by morphologic, genetic, and biochemical characters: 4T strain is a non-motile rod-shaped bacterium, with laccase activity (*34*) and slightly different plasmid profile from 4B strain (*35*), which is a motile, short, rod-shaped bacterium without laccase activity. Moreover, aged colonies of 4T strain have been shown to produce melanin-like compounds (*36*). Controls on strains were performed by microscopic examination and syringaldazide tests (*34*).

Cultures of both strains were grown in 5 ml Luria Broth. Tubes were shaken overnight at 28 °C. Bacterial cells were pelleted by centrifugation (15', 5000g), washed three times with sterile 0.7% NaCl solution, then resuspended in 2ml of sterile 0.7% NaCl solution for further assays.

Growth Medium. The effect of various electron acceptors on strain growth was investigated using a specific medium, supplemented or not by phenolic compounds. Solution 1: NH_4Cl: 1 g; K_2HPO_4: 3 g; KH_2PO_4: 2 g; malic acid: 5 g; deionized water: 948 ml; pH (by 1 M KOH) 6.8. Solution 2: $MgSO_4$, $7H_2O$: 200 mg; $CaCl_2$, $2H_2O$: 26 mg; NaCl: 100 mg ; Na_2MoO_4, $2H_2O$: 2 mg; $MnCl_2$, $4H_2O$: 7 mg; deionized water: 50 ml. Solution 3: $FeSO_4$, $7H_2O$: 631 mg; EDTA, $2H_2O$: 529 mg; deionized water: 50 ml. All solutions were autoclaved separately at 120 °C for 20 min. Solution 1 and Solution 2 were mixed and then supplemented by 1 ml of Solution 3. One ml of a filter sterilized solution of biotin (0.1 g/l) was added to the mixture to obtain a final concentration of 10^{-5} g/l. Filter sterilized phenolic compound solutions (10 mM) were added when desirable in growth flasks to a final concentration of 0.1mM.

Culture Conditions. For growth experiments, 5 x 10^{-2} ml of NaCl cell suspension were introduced in 150-ml incubation flasks filled with 30 ml of the media (see above). Two oxygen levels were used: atmospheric conditions, i. e. around 21% O_2 (21 kPa partial pressure O_2) in flasks capped with cotton plugs or 5% O_2 (5 kPa partial pressure O_2) in flasks capped with rubber stoppers. To obtain the low O_2 partial pressure, flasks filled with media were evacuated three times and filled with helium (high grade). Then 5% of the atmosphere was replaced by O_2. Finally, cell suspension and, if necessary, phenolic solutions were added. Cultivation under atmospheric conditions was performed for 48h in the dark at 28 °C under agitation. One ml subsamples of the culture medium were taken after 0h, 18h, 24h, 39h, 44h, and 48h of incubation. Cultivation under low O_2 conditions was performed in the sealed flask for 88h in the dark at 28 °C under agitation. One ml subsamples of the culture medium were taken after 0h, 22h, 44h, 48h, 64h, 69h, and 88h of incubation. Each incubation was performed in triplicate. Control incubations were conducted without phenolic compounds added. Bacterial growth was measured by optical density of cell suspensions at 540 nm.

Acetylene Reduction Activity (ARA). Plasma flasks (150 ml) were filled with 30 ml of N-free medium (growth medium described above, without NH_4Cl) and then evacuated three times and filled by helium complemented with 10% C_2H_2. The final concentration of O_2, measured by gas chromatography (catharometer) was less than 0.1%. Aliquots (1 ml) of cell suspension were introduced and flasks were incubated at 28 °C in the dark, without shaking for 120h. At this time various treatments were applied. Control: 0.3 ml of 0.7% NaCl sterile solution; CAF: 0.3 ml of 10 mM solution of CAF; O_2: 0.5% O_2 introduced in flasks; N_2O: 1.3% N_2O introduced in flasks; CAF + O_2: combination of above CAF and O_2 treatments; CAF + N_2O: combination of above CAF and N_2O treatments. Treatments were applied in triplicate. ARA was assayed at 0h, 18h, 22h, 36h, 66h, 88h, 100h, 120h, 144h, 188h, 200h,

220h, 280h, 310h, and 324h.   Ethylene in flasks was measured by gas chromatography (FID).

Redox Potential. Determination of redox potential (Eh) was performed by a platinum electrode (Ingold Messtechnik, AG Industrie) on a mV meter (CG 837, SOLEA, Lyon, France). The electrode was placed into the culture medium immediately after opening. Redox potential changed from high values soon after electrode placement, to lower values when equilibrated, and then to higher values again, when atmospheric O_2 diffused into the liquid medium. The values at the lowest point of the curve were considered as reflecting the redox potential of the medium.

Caffeic Acid Metabolite Extraction and Characterization. Thin layer (TLC) and high performance liquid (HPLC) chromatography were used for an approximate characterization of caffeic acid and its metabolites. Metabolites of CAF were extracted from the medium with ethyl acetate after a 88h incubation. The extracts were concentrated under vacuum in a rotavapor and dissolved in 0.1 ml of high grade methanol. TLC was performed on aluminium sheets coated with silica gel F254 (Merck). CAF and its metabolites were separated by migration in toluene:acetic acid (5:1). TLC plates were read at 254 and 360 nm and revealed by the diazotized benzidin reagent. HPLC was performed on a Waters 625L liquid chromatograph with a Microbondapak C18 column. First elution in gradient condition was done with a mixture of H_2O+ 2% acetic acid (A) and acetonitrile (B) (Eluent 1: flow rate 1.5 ml/min, 0-40 min A = 100%, 40-45 min A/B = 60/40, 45-50 min A = 100%). A further elution was done under isocratic conditions with pure acetonitrile (Eluent 2: flow rate 1.5 ml/min). Eluted compounds were detected by a Waters 991 photodiode array detector (Millipore).

Enzymatic Activity Measurement. Culture of 4T strain was grown in 100 ml of growth medium amended with caffeic acid (0.1 mM) at 5 kPa O_2 during one week. Bacterial cells were pelleted by centrifugation (15', 5000g), washed three times with sterile potassium phosphate buffer (50 mM, pH 7.2), then resuspended in 10 ml of the identical sterile buffer amended with caffeic acid up to a concentration of 0.1 mM. Cells were incubated at 28 °C in the dark for 4 hours. One ml subsamples of the culture medium were taken every hour, and bacterial cells were pelleted by centrifugation (15',5000g). The UV-spectrum of the supernatant of each sample was scanned in a 10 mm wide cuvette. After 4 h of incubation, bacterial cells were pelleted, washed three times, and then disrupted in 50 mM phosphate buffer at pH 7.2 by an ultrasonic treatment (Vibracell, Bioblock) at 4 °C. The cell debris was removed by centrifugation (20000g, 15', 4 °C) and supernatant was used as intracellular enzyme extract. The protein concentration of the extract was determined according to Bradford (*37*). The UV spectra of the products were obtained in a 10 mm wide cuvette containing 700 µl of phosphate buffer (amended with a polyphenolic compound to a final concentration of 0.1 mM), 100 µl of enzyme extract, and when desirable, 20 µl of 10 mM NADPH$_2$ solution in the phosphate buffer.

Results

Growth of *A. lipoferum* Strains in the Presence of Phenolic Molecules. The influence of the six phenolic compounds was investigated on the growth of strains 4B and 4T of *A. lipoferum*. Under atmospheric conditions, growth of strain 4T was inhibited by CAT and slightly lowered by 3,4-DHBA (Figure 1), while other phenolics did not influence its growth (data not shown). Under 5 kPa O_2, strain 4T

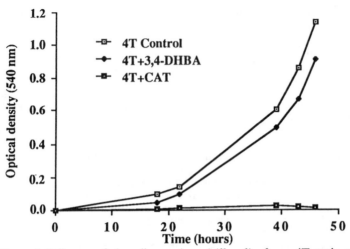

Figure 1. Influence of phenolics on *Azospirillum lipoferum* 4T strain growth under 21 kPa O_2. All standard errors of means were < 5%.

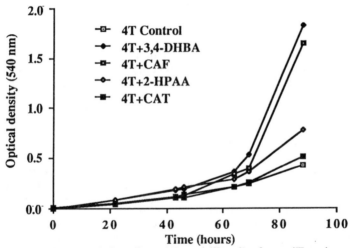

Figure 2. Influence of phenolics on *Azospirillum lipoferum* 4T strain growth under 5 kPa O_2. All standard errors of means were < 5%.

growth was stimulated by three of the six tested compounds: 3,4-DHBA > CAF > 2-HPAA and not inhibited by CAT (Figure 2). VAN and 4-HBA either slightly stimulated, or did not influence 4T strain growth (data not shown). The growth of 4B strain was not influenced by the presence of the tested phenolics under both 20 and 5 kPa partial pressure O_2 conditions (data not shown).

Caffeic Acid Metabolites and Byproducts. TLC analysis indicated (Table I) that the ethyl acetate/methanol extracts of non-incubated, sterile, fresh caffeic acid solution was eluted as two spots (two isomers) exhibiting bright blue fluorescence.

Extracts from caffeic acid incubated under sterile conditions under both 20 and 5 kPa O_2 were eluted as two spots with higher Rf values showing a much more lipophilic character than components from the fresh caffeic acid solution (degradation products). Incubation under atmospheric O_2 with both strains resulted in a one spot chromatogram with another Rf value. This spot exhibited a weaker blue/yellow fluorescence when compared to the higher Rf spot of fresh caffeic acid. Under 5 kPa O_2 conditions, contrasting results were obtained from incubation of caffeic acid with strains 4B and 4T. Under these conditions, methanol extracts of caffeic acid incubated with 4B strain resulted in one spot with the lowest Rf of all the TLC runs. With 4T strain, three spots were obtained: the compound with the less lipophilic nature (lower Rf) and a bright blue fluorescence was close to the more lipophilic component of fresh caffeic acid but it did not persist after 24h; the component with the more lipophilic Rf was close to the more lipophilic component from sterile caffeic acid incubation under limited 5 kPa O_2; and a new spot appeared in the intermediate zone of Rf.

High Perfomance Liquid Chromatography. HPLC confirmed and refined most of the TLC results. Results from HPLC analyses are summarized in Table II. Fresh caffeic acid HPLC resulted in two peaks (two spots in TLC) eluted at 16.75 and 17.80 min. respectively, with equivalent O.D. at 237 and 323 nm. No peak eluted in acetonitrile. Products of caffeic acid incubation in sterile conditions either under atmospheric conditions or under 5 kPa O_2 were eluted by acetonitrile only as one peak after 1.4 min. and 3.4 min. respectively, when traces of less lipophilic components

Table I. **Thin layer chromatography of the products resulting from the incubation of strains 4T and 4B of *A. lipoferum* with caffeic acid**

Treatment	Spots number	100 Rf	Color of fluorescence	Stability
CAF (fresh solution)	1 2	5 20	bright blue bright blue	stable stable
CAF atm. condit.	1 2	89 72	blue blue	stable stable
CAF (5 kPa O_2)	1 2	87 70	blue blue	stable stable
4B + CAF atm. condit.	1	17	weak blue/yellow	stable
4T + CAF atm. condit.	1	17	weak blue/yellow	stable
4B + CAF (5 kPa O_2)	1	4	blue/brown	stable
4T + CAF (5 kPa O_2)	1 2 3	22 57 87	bright blue blue blue	unstable stable stable

Table II. HPLC of major products resulting from the incubation of strains 4T and 4B of A. lipoferum with caffeic acid

Treatment	Elution by Eluent 1			Elution by Eluent 2		
	Peak nb	λmax (O.D.)	Retention time (min)	Peak nb	λmax (O.D.)	Retention time (min)
CAF (fresh solution)	1	237 (0.18) 323 (0.30)	16.75	none		
	2	237 (0.21) 323 (0.35)	17.80	none		
4B + CAF (20 kPa O$_2$)	1	234 (0.41) 323 (0.64)	18.16	none		
4T + CAF (20 kPa O$_2$)	1	237 (0.11) 323 (0.18)	17.96	none		
4B + CAF (5 kPa O$_2$)	1	225 (0.14) 345 (0.11)	17.35	none		
	2	234 (traces) 327 (traces)	17.96	none		
4T + CAF (5 kPa O$_2$)	1	236 (0.11) 323 (0.18)	17.92	1	247 (0.6) 334 (0.35)	2.12
	2	230 (0.18)	22.82	2	281 (0.1)	2.4
				3	280 (0.1)	3.4
CAF (20 kPa O$_2$)	none			1	272 (0.18)	1.4
CAF (5 kPa O$_2$)	none			1	280 (0.24)	3.4

were eluted in isocratic conditions. Incubation of caffeic acid with strain 4B under atmospheric conditions resulted in one peak eluted at 18.16 min (i.e. 20 sec later than the second peak of fresh solution of CAF) with O. D. maxima at 234 and 323 nm, each representing the sum of the optical densities of the corresponding two peaks of the fresh caffeic acid solution. Identical incubation conditions with strain 4T resulted in the same elution time (17.96 min) with O. D. maxima at 237 and 323 nm, consistently lower than with strain 4B.

Under low O$_2$ conditions, strains 4B and 4T gave completely different HPLC patterns of caffeic acid biotransformation. Strain 4B products were represented by only traces of the 234-327 nm absorbing substance at 17.96 min and by a main peak absorbing at 225 and 345 nm which eluted at 17.35 min, i.e. a weakly lipophilic product (as in TLC run). Strain 4T products were partly eluted during the isocratic phase of HPLC with two main peaks at 17.92 min with maximum O. D. at 236 and 323 nm (i.e. very close to the more lipophilic product from caffeic acid fresh solution) and at 22.82 min with a maximum absorbance at 230 nm. This last peak corresponded to a new compound unrelated to any other caffeic acid product (the intermediate Rf spot in TLC assay). Three other products eluted in pure acetonitrile. The main new component was recovered after 2.12 min and strongly absorbed at 247 and 334 nm. The third component exibited similar HPLC properties as the product of sterile caffeic acid incubated in 5 kPa O_2.

Nitrogenase Activity and Redox Potential. Nitrogenase activity was assayed through Acetylene Reduction Activity (ARA). Strain 4T of *A. lipoferum* only will be considered because of its specific reaction when polyphenolics were added, and because ARA of strain 4B was not influenced by the addition of caffeic acid in any combinations (data not shown). As described in Material and Methods, ARA of strain 4T was first assayed in the N-free growth medium without electron acceptor added during 120h. After a lag phase (Figure 3), C_2H_4 production increased up to 27 nmol per hour in all flasks, continued until day 3, and then stopped completely, due to oxygen depletion.

At 120h, the introduction of electron acceptors in treated samples or of NaCl solution in control flasks, slightly stimulated ARA. The amount of C_2H_4 produced ranged from 50 to 107 nmole per flask, i.e. approximately 2 to 4 nmole per hour per flask. During the following hours no additional ARA was detected. Forty-four hours after introduction (i.e. at 164 after the experiment began), ARA started again in O_2, O_2 + CAF (Figure 4), and N_2O + CAF treatments (Figure 3) and continued during the following 86h until day 10. ARA in control flasks, N_2O, and CAF treaments remained at a basic low level (Figure 3). At 300h, ARA had ceased in all samples.

During the first 120 hours, the redox potential decreased from 255 mV to 94 mV (Figure 5). The introduction of O_2 or O_2 + CAF increased redox potential of the medium to 120 mV (Figure 4). Redox potential of the samples treated with other electron acceptors was unchanged (Δ mV = 4 to 6 mV). In subsequent measurements, the redox potential of control samples was further decreasing. At the same time, the

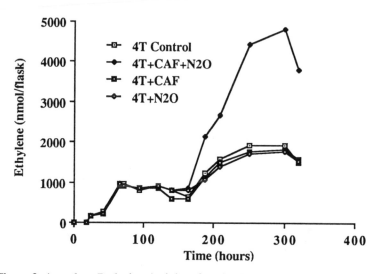

Figure 3. Acetylene Reducing Activity of strain 4T of *A. lipoferum* under electron acceptor-limiting conditions. Caffeic acid and N_2O were added to the media at time 120. Standard errors of means were < 4% in all cases.

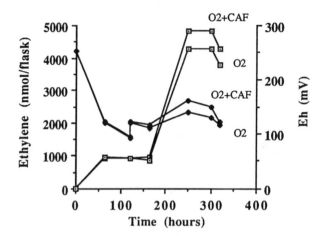

Figure 4. Acetylene Reducing Activity (open symbols) and redox potential after addition of O_2 and CAF+O_2 (at time 120) to strain 4T of *A. lipoferum*. Standard errors of means were < 4% in all cases.

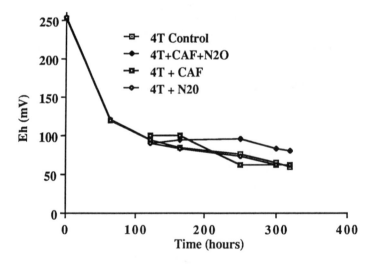

Figure 5. Redox potential of the medium when strain 4T of *A. lipoferum* was grown under electron acceptor-limiting conditions. Caffeic acid and N_2O were added to the media at time 120.

redox potential of N_2O + CAF remained unchanged up to 250h and then slowly decreased (Figure 5). The difference between N_2O + CAF and control samples at the end of the experiment was around 20 mV. Redox potential of the medium supplemented by CAF was only slightly higher than the control during 44h and then rapidly decreased to a level lower than the control. In O_2 and O_2 + CAF treatments, O_2 diffusion was apparently slow; maximum redox potential was not reached 2 days after addition of O_2. Maximum recorded values were 140 (O_2) and 160 (O_2 + CAF) mV respectively at time 250h (Figure 4).

Enzymatic Activity of Cells and Cell-free Extracts of *A. lipoferum* 4T.
No spontaneous changes in UV spectra of CAF, 3,4-DHBA, and CAT were ever observed when these compounds were incubated in potassium buffer only (Table III). No changes in UV-spectra of CAT were observed in any treatment: fresh solution, CAT + 4T cell-free extract, and CAT + 4T cell-free extract + NADPH$_2$. During incubation of 4T cells with caffeic acid in potassium phosphate buffer, the absorbance peak of the supernatant shifted from 285 nm to 269 nm after 4h of incubation. 4T cell-free extract (1 mg/ml of proteins) gave a comparable shift under 5 kPa O_2 only when NADPH$_2$ was provided along with caffeic acid. This shift of the absorbance peak was 7 nm (from 285 to 278) after 35 min of incubation. At the same time, we observed a 50% decrease in the optical densities at 285 and 310 nm (data notshown), i.e. a 50% decrease in caffeic acid concentration within 35 min. In the absence of NADPH$_2$, the

Table III. Changes in UV-spectra of phenolics after reaction with *A.lipoferum* 4T cells or cell-free extract

Treatment	Compound	Reaction time (min.)	λ max.
Fresh solution	CAF	-	285; 310
4T strain, cells	CAF	240	269; 310
4T strain, extract	CAF	35	285
4T strain, extract +NADPH$_2$	CAF	35	278
NADPH$_2$	CAF	35	285; 310
Fresh solution	CAF	35	285; 310
Fresh solution	3,4-DHBA	-	244; 286
4T strain, extract	3,4-DHBA	35	244; 286
4T strain, extract +NADPH$_2$	3,4-DHBA	35	241; 278
NADPH$_2$	3,4-DHBA	35	244; 286
Fresh solution	3,4-DHBA	35	244; 286
Fresh solution	CAT	-	279
4T strain, extract	CAT	35	279
4T strain, extract +NADPH$_2$	CAT	35	279

decrease in concentration of caffeic acid was more pronounced, but no shift of major peaks was registered. A solution of CAF receiving only $NADPH_2$ was indistinguishable from a fresh solution fo CAF alone.

The wavelength shift of the absorbance peak, due to 35 min incubation of the cell-free extract of 4T with 3,4-DHBA was nearly the same as that for caffeic acid. There was no decrease in 3,4-DHBA absorbance, regardless of the presence or the absence of $NADPH_2$ in the mixture.

Discussion

Motile vs Non-Motile *Azospirillum*. The role of phenolic compounds on growth and nitrogenase activity was investigated in *Azospirillum lipoferum*, a common rhizosphere inhabitant. Strains 4B and 4T have been isolated from the rhizosphere dominant nitrogen fixing flora of field grown rice and are thus supposed to represent bacteria well adapted to the presence of a grass host plant. They were selected for their contrasting properties under low oxygen partial pressures, and their seemingly peculiar phenolic metabolism. In previous studies, it had been shown that only the non-motile 4T strain could exibit a high level of nitrogenase activity without growth under 1 kPa O_2 partial pressure (*20*). Furthermore, it was later shown that this non-motile strain possessed a unique laccase activity (*34*) and was able to produce melanin through polyphenol condensative oxidation (*36*).

Phenolics. The phenolics studied here were chosen as representatives of the various molecular structures of polyphenolic compounds likely to be encountered in rhizospheric environments, either because they are part of root exudates or because they are intermediates in the biodegradation of molecules containing aromatic rings (lignin, tanins of plant origin, xenobiotics from agricultural sources) (*38, 5*). All these phenolic compounds are able to undergo redox transformation: they can exist as free radicals with an unpaired electron and thus can donate or accept electrons.

Effects on Growth. Under high oxygen partial pressure, the addition of 0.1 mM phenolics to the growth medium of strain 4B produced no measurable effects. The carbon added represented only 0.1% of the carbon provided by malate in the growth medium. Nevertheless, low concentrations of phenolics had a measurable effect on strain 4T: a slight decrease in growth rate for 3,4-DHBA, and a strong growth inhibition with catechol. Orthoquinone is well known for its interaction with amino acids (*29*), and subsequent inhibition of enzymatic activities (*39*), but the reason for a strain 4T specific inhibitory effect of catechol on growth remains unclear. Under 5 kPa O_2, 4T growth was strongly stimulated by 3,4-DHBA and CAF, slightly increased by 2-HPAA, and unaffected by CAT. The contrasting effect of 3,4-DHBA, CAF, and CAT on the growth of strain 4T under 5 kPa O_2 and 21 kPa O_2, with the concomitant appearance of new lipophilic products under low O_2, raised the hypothesis of a relationship between respiration and the transformation of phenolics. Under the conditions of our experiment, each polyphenolic compound had at least one site where an electron could be transferred. The fact that 3,4-DHBA and CAF similarly affected strain 4T growth supported the opinion that the ethylenic group of the side chain of caffeic acid was not involved in electron transfer. Catechol toxicity probably balanced the possible beneficial electron acceptor role of the oxidized counterpart (orthoquinone) of this molecule.

Metabolites. The pattern of byproducts observed by TLC and HPLC illustrates the complexity of CAF metabolism. TLC and HPLC data revealed that strain 4B limited the transformation of CAF to the most probable dimer under atmospheric conditions, and to a lesser extent, an unknown hydrophilic byproduct under low O_2. On the contrary, the non-motile strain 4T strongly reacted with caffeic acid under both pO_2. Under atmospheric conditions, strain 4T probably utilized caffeic acid as an additional source of carbon and energy, because HPLC revealed only 1/4 of the initial absorbance of caffeic acid without the appearance of other compounds. Under 5 kPa O_2, highly lipophilic byproducts (Table 1) appeared, as well as an unstable intermediate phenolic compound.

Acetylene Reduction. The hypothesis about a relationship between respiration and transformation of phenolics suggested the study of acetylene reduction instead of growth since nitrogenase activity (ARA) requires large amounts of ATP and since ATP in *Azospirillum* is essentially generated by the electron respiratory transfer chain. It was thus easy to obtain conditions under which the amount of ethylene evolved was directly correlated to the amount of terminal electron acceptors available. These electron acceptor-limiting conditions were obtained when available oxygen had been consumed in the closed vials, i.e. between 36 and 66 hours (Figure 3). At that stage, the addition of O_2 (Figure 4) triggered a new phase of ARA in the O_2 and O_2 + CAF treatments, confirming that the amount of electron acceptor was the limiting factor of nitrogenaese activity and that enough carbon and energy had been supplied by malate in the medium.

It is likely that the slight increase of ARA following the addition of NaCl (in control) and N_2O or CAF in the other assays at 120h was due to the concomitant introduction of traces of O_2. The corresponding ethylene evolution (1 micromole per flask) was quite low compared to the 4 micromoles obtained when 0.5 kPa partial pressure of oxygen was present. In contrast, the co-introduction of O_2 or caffeic acid plus re-oxidizing agents (N_2O) resulted in maximal ARA stimulation. In this treatment, the concentration of CAF was too low, compared to malate, to play any role as a source of carbon. Moreover, the only possible role for N_2O was to oxidize caffeic acid: it could not be used as a respiratory terminal electron acceptor due to the presence of acetylene. It is thus suggested that an oxidized derivative of CAF was able to accept the electrons from *Azospirillum* metabolism.

The stimulation of ARA by the addition of CAF in the presence of 0.5 kPa O_2 was around 25% more than stimulation by 0.5 kPa O_2 alone (Figure 4). This suggests that the electron acceptor for ARA was predominantly O_2 and secondarily the oxidized derivative of caffeic acid.

An unclear point is the actual reoxidizing agent: is N_2O oxidizing CAF directly through the oxidation of another intermediary compound, such as Fe^{+2} (*28*). This question is currently under investigation.

Redox Potential Changes. The low increase in redox potential of the medium immediately after the addition of CAF alone was probably due to the existence of a partly oxidized form in the solution at the time of introduction (Figure 5). The redox potential of the medium supplied N_2O + CAF was unchanged up to 250h of incubation, and the difference of redox values between this treatment and the control was around 20 mV at the end of the experiment. The maximal increase of redox potential was observed in O_2 + CAF treatment and the maximal difference between this treatment and treatment by O_2 only was around 20 mV as well (Figure 4.). Therefore, a 20 mV increase in redox potential resulted from the presence of caffeic acid, which is consistent with the results obtained earlier (*40*) and would demonstrate the presence of an oxidized form of the phenolic molecule.

The Electron Acceptor. The interaction of the cells and the cell-free extract of strain 4T with polyphenolics showed a significant shift of maximal absorbance for CAF and 3,4-DHBA. The same type of shift can be seen during the formation of the reduced configuration of polyphenolic molecule from the oxidized form by electron transfer (*31*). The absence of this shift in the absence of cells or cell-free extracts indicates that the transfer of electrons to the oxidized conformation of polyphenolic compounds is a biological process brought about by strain 4T. In cell-free extract, $NADPH_2$ could serve as a suitable electron donor. This phenomenon is somewhat substrate specific as no interaction was observed with CAT. The actual electron acceptor should be one of the oxidized derivatives of polyphenolics, either the semiquinone or the free radical with an unpaired electron, or products of quinone self-condensation. This hypothesis needs further work to be elucidated. Currently we are studying the structures of the unstable intermediate and of the lipophilic compound produced by strain 4T from CAF.

Conclusion

The microbial degradation of phenolics through their utilization as a source of carbon is well documented. The capacity of humic acids and phenolic compounds to act chemically as electron donors or acceptors has been recognized for a long time (*41, 42*), as well as the electron-transfer properties of phenolics which interfere with the functioning of enzymes (*9*). The ability of phenolics to stimulate the respiration of soil microorganisms has also been reported (*43-45*), as well as the ability of phenoxy acid herbicides to stimulate the nitrogenase activity of root-associated *Azospirillum* (*46*). Nevertheless, to our knowledge, it has never been suggested that this stimulation could proceed through an enzymatic electron transfer system, allowing the microaerophilic bacterium, in the absence of oxygen or under low oxygen, to use phenolic compounds as a respiratory terminal electron acceptor.

   This hypothesis about the capacity of polyphenolics to be used as an electron acceptor by a non-motile *Azospirillum lipoferum* strain is coincidental with the discovery that such strains also have the capacity to achieve the oxidative transformation of phenolic compounds (*34*). This suggests that the adaptation of *Azospirillum* to the rice rhizosphere, could involve the use of phenolic redox couples to refine the tuning of the bacterial environment to a redox value optimal for the functioning of its nitrogenase.

   Such transformations of phenolics are likely to be accompanied by changes in their degee of polymerization, solubility, lipophilicity, and could thus significantly contribute to their fate in soils, especially in the rhizosphere. Soil organic matter turnover along the roots could thus eventually differ largely from bulk soil due to the activity of some specialized bacteria comparable to strain 4T. Xenobiotic compounds (including haloaromatics) could be processed along the same lines in soils, provided they share enough homology with naturally occuring aromatic compounds. This could constitute the basic rationale for the use of active rhizospheres to harness the transformation of some xenobiotics in soils.

Acnowledgments

This study was supported by a fellowship awarded to A. B. by the Centre National de la Recherche Scientifique, Paris (France). The authors thank Dr. R. Lensi for N_2O suggestion and facilities and Dr. R. Bally for helpful scientific assistance.

Literature Cited

1. Hsu, T. S.; Bartha, R. *Appl. Environ. Microbiol.* **1979**, *37*, 36-41.
2. Sandmann, E.; Loos, M. A. *Chemosphere.* **1984**, *13*, 1073-1084.
3. Rasolomanana, J. L.; Balandreau, J. *Rev. Ecol. Biol. Sci.* **1987**, *24*, 443-457.
4. Walton, B. T.; Anderson, T. A. *Appl. Environ. Microbiol.* **1990**, *54*, 1012-1016.
5. Blum, U.; Wentworth, T. R.; Klein, K.; Worsham, A. D.; King, L. D.; Geric, T. M.; Lyu, S. W. *J. Chemical Ecol.* **1991**, *17*, 1045-1068.
6. Guenter, F. R.; Parris, R. M.; Chesler, S. M.; Hilpert, L. R. *J. Chromatogr.* **1981**, *207*, 256-261.
7. Murakami, S.; Nakanishi, Y.; Shinke, R.; Aoki, K. *Soil Biol. Biochem.* **1991**, *23*, 815-819.
8. Hussein, Y. A.; Tewfic, M. S.; Hamdi, Y. A. *Soil Biol. Biochem.* **1974**, *6*, 377-388.
9. Chen, Y. P.; Glenn, A. R.; Dilworth, M. J. *FEMS Microbiol. Lett.* **1984**, *21*, 201-205.
10. Gajendiran, N.; Mahadevan, A. *Indian J. Exp. Biol.* **1990**, *28*, 1136-1140.
11. Barkovskii, A. L. *Effluent Treatment and Waste Minimization*; IChem Symp. Ser., 132; Chameleon Press: London, UK, 1993; pp 157-169.
12. Okon, Y.; Kapulnik, Y. *Plant & Soil.* **1986**, *90*, 9-16.
13. Levanony, H.; Bashan, Y. *Plant & Soil.* **1991**, *137*, 91-97.
14. Fages, J.; Arsac, J. *Plant & Soil,* **1991**, *137*, 87-90.
15. Fages, J. *Symbiosis.* **1992**, *13*, 15-26.
16. Barkovskii, A. L.; Shub, G. M. *Mikrobiologiya,* **1986**, *55*, 237-240.
17. Dagley, S. In *Soil Biochemistry*; McLaren, A. D.; Peterson, G. H., Eds.; Marcel Dekker, Inc: New York, 1967; pp 287-317.
18. Karasevitch, Y. N. *The foundation of selection for microorganims utilizing synthetic organic compounds*; Mir: Moscow, RU, 1982.
19. Korzhenevitch, V. I.; Ignatov, 0. V.; Mironov, A. D.; Krivopalov, Y. V.; Barkovskii, A. L. *Prikladnaja Biochimiya i Microbiologiya.* **1991**, *27*, 365-369.
20. Heulin, T.; Weinhard, P.; Balandreau, J. In *Azospirillum II, Genetics, Physiology, Ecology. Experientia Suppl.*; Klingmüller, W., Ed.; Birkhäuser Verlag: Basel, CH, 1983, 48, pp 89-94.
21. Ueckert, J.; Hurek, T.; Fendrik, I.; Niemann, E.-G. *Plant & Soil.* **1992**, *122*, 59-65.
22. Flessa, A.; Fischer, W. R. *Plant & Soil.* **1992**, *143*, 55-60.
23. Lynch J. M.; Whipps, J. M. *Plant & Soil.* **1990**, *129*, 1-10.
24. Havelka, U. D.; Boyle, M. G.; Hardy, R. W. F. In *Nitrogen in Agricultural Soils*; Agronomy Series; Stevenson, F. J., Ed.; ASA, CSA and SSSA: Madison, Wisconsin, 1982, Vol.22; pp 365-422.
25. Crane, F. L.; Barr, R. In *Coenzyme Q. Biochemistry, Bioenergetics and Clinical Applications of Ubiquinone*; Lenaz, G. Ed.; Wiley & Sons: New York, 1985, 1-39.
26. Gasser, F.; Biville, F; Turline, G. *Ann. Inst. Pasteur/Actualites.* **1991**, *2*, 139-149.
27. Lehmann, R. G.; Cheng, H. H.; Harsh, J. B. *Soil Sci. Soc. Am. J.* **1987**, *51*, 352-356.
28. Deiana, S.; Gessa, C.; Manunza, B.; Marchetti, M.; Usai, M. *Plant & Soil.* **1992**, *145*, 287-294.
29. Flaig, W.; Beutelspacher, H.; Rietz, E. In *Soil Components, Organic Components*; Gieseking, J. E., Ed.; Springer Verlag: Berlin, 1975, pp 1-211.

30. Lehmann, R. G.; Cheng, H. H. *Soil Sci. Soc. Am. J.* **1988**, *52*, 1304-1309.
31. Swallow, A. J. In *Function of Quinones in Energy Conserving Systems*; Trumpower, B. L., Ed.; Academic Press: London, UK, 1982, pp 59-73.
32. Yoshinari, T.; Hynes, R.; Knowles, R. *Soil Biol. Biochem.* **1977**, *9*, 177-183.
33. Thomas-Bauzon, D.; Weinhard, P.; Villecourt, P.; Balandreau, J. *Can. J. Microbiol.* **1982**, *28*, 922-928.
34. Givaudan, A.; Effosse, A.; Faure, D.; Potier, P.; Bouillant, M. L.; Bally, R. *FEMS Microbiol. Lett.* **1993**, *108*, 205-210.
35. Bally, R.; Thomas-Bauzon, D.; Heulin, T.; Balandreau, J.; Richard, C.; De Ley, J. *Can. J. Microbiol.* **1983**, *29*, 881-887.
36. Givaudan, A.; Effosse, A.; Bally, R. In *Nitrogen Fixation. Developments in Plant and Soil Sciences*; Polsinelli, M.; Materassi, R.; Vincenzini, M, Eds.; Kluwer Academic: Dordrecht, 1991, pp 311-312.
37. Bradford, M. *Anal. Biochem.* **1976**, *72*, 248-254.
38. Gross, D. *Z. Pflanzenkrankh. Pflanzensch.* **1989**, *96*, 535-553.
39. Pflug, W.; Ziechmann, W. *Soil Biol. Biochem.* **1981**, *13*, 239-299.
40. Einarsdottir, G. H.; Stankovich, M. T.; Tu, S-C. *Biochemistry.* **1988**, *27*, 3277-3285.
41. Ziechmann, W. von. *Geoderma.* **1972**, *8*, 111-131.
42. Ziechmann, W. von. *Z. Pflanzenernahr. Bodenkd.* **1977**, *140*, 133-150.
43. Krotzky, A.; Berggold, R.; Jaeger, D.; Dart, P. J.; Werner, D. *Z. Pflanzenernahr. Bodenkd.* **1983**, *146*, 634-642.
44. Spaling, G. P.; Ord, B. G.; Vaughan, D. *Soil Biol. Biochem.* **1981**, *13*, 455-460.
45. Werner, D.; Krotzky, A.; Berggold, R.; Thierfelder, H.; Preiss, M. *Arch. Microbiol.* **1982**, *132*, 51-56.
46. Haahtela, K.; Kilpi, S.; Kari, K. *FEMS Microbiol. Ecol.* **1988**, *53*, 123-127.

RECEIVED February 15, 1994

Chapter 4

Effect of Nutrients on Interaction Between Pesticide Pentachlorophenol and Microorganisms in Soil

Kyo Sato

Institute of Genetic Ecology, Tohoku University, 2−1−1, Katahir, Aoba-ku, Sendai, 980 Japan

An investigation was conducted on the interation between the pesticide PCP (pentachlorophenol) and bacteria in soil supplemented with amino acids and carbohydrates in light of the fact that nutrient concentrations for microorganisms are greater in rhizosphere soil than in non-rhizosphere soil. PCP did not markedly delay the start of degradation of the amino acids and glucose, and the numbers of bacteria also increased in connection with the degradation. Bacterial flora established by PCP application were modified by the addition of amino acids. Degradation of PCP, and the increase in the number of PCP-degrader(s) were delayed in the presence of amino acids and glucose, but not in the presence of cellulose. These results suggest that, in general, the interaction between pesticides and microorganisms in rhizosphere soil is different than that in non-rhizosphere soil.

Pentachlorophenol (PCP) has been used as both a herbicide and a fungicide. In Japan, its application to paddy rice fields as a herbicide is prohibited because of its toxicity to organisms living in the water. However, PCP is still used as a fungicide for upland crops or timber.

Many pesticides introduced into soil environments affect soil microorganisms, although most pesticides are known to be degraded by microorganisms in soil. Therefore, many studies have been conducted on the interrelationships between pesticides and microorganisms in soil.

The behavior of microorganisms in soil is influenced by the presence of microbial nutrients, and thus, the composition of soil microflora is regulated by the types and amounts of nutrients. The nutrients may also be easily utilized by pesticide-degrading microorganisms in soil, as many of the pesticide-degrading microorganisms isolated so far are common species not nutritionally fastidious (1,2). Therefore, it is very important to know how microbial nutrients modify the composition of the microflora established by the impact of pesticide application, or how the behaviors of pesticide-degrading microorganisms are affected by the nutrients. This situation is especially relevant to the rhizosphere because plant roots exude various kinds of microbial nutrients.

In this report, the effects of some microbial nutrients added to soil on both soil bacterial flora established by PCP, and microbial degradation of PCP are presented.

0097−6156/94/0563−0043$08.00/0

Materials and Methods

Percolated Soils. Methods of soil-percolation, counting the numbrs of bacteria in the percolated soils, as well as chemical analysis of PCP and several nitrogen compounds in the percolates were previously reported (*3,4*). Isolation of bacteria from the percolated soils and identification of the isolated bacteria were also reported (*5*).

PCP-Tolerant and/or Degrading Bacteria. Ten grams of soil suspended in 200 ml mineral salts solution containing ammonium nitrate was incubated in a reciprocal shaker at 22 °C in the dark. PCP was added to the suspended soil at several concentrations. Samples of the suspended soil were withdrawn at appropriate intervals and submitted to bacterial enumeration and chemical analyses. PCP-tolerant bacteria were enumerated by the dilution agar plate method using albumin agar medium treated with 100 ppm PCP. Populations of PCP-degading bacteria was estimated by the most probable number (MPN) technique. Basal medium (*1*) was supplemeted with 0.01% yeast extract and 5 ppm PCP. Five tubes were inoculated per serial dilution of the samples. After 5 weeks incubation at 22 °C, the presence of PCP in the tubes was checked by an absorption peak near 320nm. If the absorption peak disappeared, the test tube was considered positive for growth of PCP-degrading bacteria. PCP in soil suspensions was analyzed similarly to that in the percolate.

Effect of Nutrients on Degradation of PCP. A shaking soil suspension containing several concentrations of amino acids, glucose, or cellulose (powder) was treated with 10 ppm PCP. PCP-degrading bacteria were enumerated by the same procedure as described above. Amino acids and glucose were analyzed colorimetrically using ninhydrin and anthrone, respectively. Carbon dioxide evolved from soil was absorbed in 0.5 N NaOH. Periodically, titrations were carried out for the unneutralized NaOH with 0.5 N HCl.

Results and Discussion

Effect of PCP on Soil Bacterial Flora in Percolated Soils

Changes in Population of Bacteria in Soil Treated with Glycine. Total viable counts increased with the length of incubation in all plots, although the increase was retarded slightly in the plot treated with PCP. The numbers attained were similar in the plots with and without PCP, while those in the plot without the addition of glycine were smaller to those in plots with glycine (Figure 1) (*3*). PCP in the plot did not dissipate during the experimental period (*3*). These results indicate that the bacteria were not substantially influenced by the presence of the pesticide. The pattern of the changes in total viable counts was very simlar to that of gram-negative bacteria (*3*), whose populations were present at very low levels at the initial time of incubation and increased dramatically soon after incubation. This suggests that the predominant bacteria in the plots treated with glycine are gram-negative.

The effect of PCP on populations of bacteria was also tested at different PCP concentrations (Figure 2) (*6*). The maximum number of viable bacteria attained in the respective plots was quite similar regardless of the difference in the amount of PCP, although some modifications were observed in the pattern of population changes among the plots: 200 ppm PCP delayed the initial increase, and after a short period at

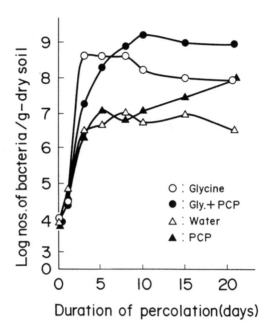

Figure 1. Changes in the population of total viable bacteria in the percolated soils(*3*).

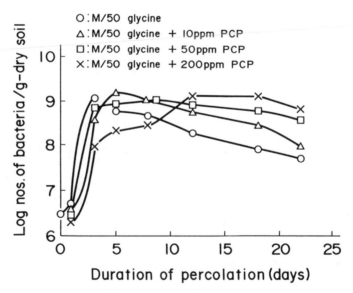

Figure 2. Changes in the population of total viable bacteria in percolated soils treated with different concentrations of PCP(*6*).

the maximum level, the numbers decreased slowly in the soils amended with PCP, while they decreased soon after reaching maximum numbers in the soil without the addition of PCP. This suggested that the larger the amount of PCP applied, the longer the maximum level was maintained and the more slowly the counts decreased. PCP did not dissipate in this case, and the population changes of the total viable count were also very similar to the gram-negative counts (6). These results indicate that PCP did not affect the changes in the population of bacteria in the soil.

Changes in Populations of Bacteria in Soil Treated with Amino Acids other than Glycine. For generalizing the effect of PCP on the bacterial population, the pattern of changes in the population was followed in soil treated with amino acids other than glycine. Glutamate, leucine, or phenylalanine was added to the soil. For all amino acids tested, the total numbers in the respective soils increased very similarly between the plots with and without the addition of PCP, although the initial increase started more slowly and the maximum numbers attained persisted longer in the plots with PCP than in the plots without PCP (Figure 3) (4). The changes in numbers of total viable bacteria progressed very similarly to the numbers of gram-negative bacteria. This phenomenon was similar to results of glycine addition. Amino acids in soil decreased with time of incubation irrespective of PCP application (4). PCP did not dissipate in soils during the experiment period. These results also show that the bacterial population in soil is apparently not affected by the application of PCP under different nutrient conditions.

In all cases, the populations attained decreased more slowly in the soils treated with PCP than in those without PCP. PCP is biocidal and may kill some microorganisms in soil. The direct correlation between the persistence of the maximum bacterial numbers and the amount of PCP added can be partly explained by "the partial sterilization effect", as reported for fumigation or solvent treatment of soil (7). However, other factors may also be involved, as maximum numbers were quite similar regardless of the presence or absence of PCP.

Microbiological Processes in Soil Treated with Glycine. In the preceeding sections the populations of several bacterial groups were observed to be unaffected by PCP application under several conditions. In all experiments mentioned above, the amino acids applied were degaded. In general, amino acids may be transformed to nitrate via ammonium and nitrite under the aerobic conditions adopted in these experiments. All of these processes are known to be carried out microbially, and different microorganisms are involved in the respective reaction steps. The degradation of glycine as affected by the addition of PCP is shown in Figure 4 (3). PCP appears to retard the decomposition of glycine, and this retardation was enhanced by the soluble form of PCP, Na-PCP. This probably indiactes that the soluble form affected all microorganisms, being distributed evenly into the soil, while the unsoluble form of PCP was not so easily distributed into all parts of the soil as to affect all microorganisms. At any rate, glycine disappeared rapidly after incubation, and this disappearance contributed to the proliferation of bacteria in both the presence and the absence of PCP, and also to ammonium production.

Nitrification, on the other hand, was greatly inhibited by PCP: ammoniification was successively followed by nitrification in the soil with the addition of only glycine, while nitrification started after a long lag in the soil treated with PCP. Complete inhibition was observed in the case of Na-PCP addition, indicating a good distribution of Na-PCP into the soil.

Based on the results observed, it is concluded that PCP does not greatly affect heterotrophic processes such as ammonification of glycine. The total viable counts of bacteria, which increased similarly regardless of PCP application in the soil treated with glycine, may be primarily responsible for the heterotrophic process.

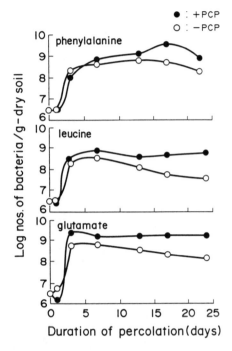

Figure 3. Changes in the population of total viable bacteria in percolated soils with the addition of different amino acids(*4*).

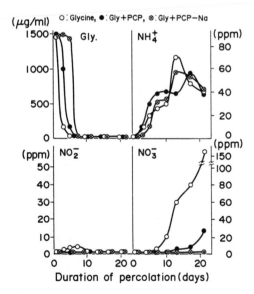

Figure 4. Changes in the amounts of glycine, ammonium, nitrite, and nitrate in the leachates(*3*).

Bacterial Flora in Soil Treated with Glycine. Studies on the affect of pesticides on soil microorganisms have been carried out, from the perspective of soil biochemistry, measuring microbial activities such as CO_2 evolution, ammonification, and nitrification (8,9). Some research has been reported on the quantitative microbial changes at one time after the addition of peticides (10,11), but there are few studies on successive quantitative microbial changes (12), and follow-up surveys of specific microbial groups at the genetic level are very scarce (13). In the preceeding section, the same pattern of glycine degradation, accompanied by a successive increase in numbers of bacteria, was reported in glycine-percolated soil whether PCP was added or not. This seems to indicate that PCP does not affect either bacteria or the degradation of glycine in the soil. The present section describes the confirmation of the effect of PCP on bacterial flora in the glycine-percolated soils. For this purpose, the bacterial strains isolated from several plots of the percolated soil were assorted taxonomically.

Since a great majority of the isolated strains were gram-negative, a differentiation was focussed on the gram-negative bacteria. Figure 5 indicates that the bacterial flora varies among the different percolated soils (5). Many strains isolated from the percolated soils with the treatment of PCP died within 15 days after the purified cultures were stored. Thus, these strains could not be bacteriologically characterized. Bacterial flora differed markedly between glycine- and water-percolated soils. One group predominated in the former soil (group VIIc, over 50%), while various groups were equally isolated from the latter soil. A similar pattern of bacterial flora in the percolated soils was also observed in a previous study (14).

The surviving bacterial groups isolated from the percolated soils treated with PCP varied with the soils. They also differed from those in both glycine- and water-percolated soil without PCP treatment. One group (VIId), which predominated in the glycine-percolated soil with PCP treatment differed from one group (VIIc) which predominated in the glycine-percolated soil without PCP treatment. The bacterial flora in the PCP-percolated soil with glycine was more diverse compared with that in the PCP-percolated soil without glycine.

These results suggest that the composition of bacterial flora in the glycine-percolated soil was modified by the addition of PCP even though the population appeared not to be affected. Furthermore, PCP modified the bacterial community such that only a few specific groups of bacteria were able to grow. This modification was affected by soil conditions such as nutrients available for microorganisms.

The bacterial group predominating in the PCP-percolated soil without glycine had the highest PCP-tolerance. The addition of glycine modified the bacterial flora by weakening the tolerance of the bacteria (15). This may relate to the restriction of growth of various groups of bacteria in this soil. The restriction was alleviated by the addition of nutrients, suggesting neutralization of the impact of pesticide application by microbial nutrients. The above results indicate that qualitative analysis of microorganisms is important for environmental assessment of pesticide application.

Increase in Numbers of PCP-Tolerant and/or Degrading Bacteria in Soil

PCP-Tolerant Bacteria. Although the bacterial populations increased with time of incubation regardless of the addition of PCP, the pattern of the increase differed among the soils treated with different amounts of PCP. The initial increase was slowed in the plots treated with larger amounts of PCP compared with those treated with smaller amounts of PCP. However, the population in the former attained a higher level than the latter after a few weeks of incubation. This suggests that the increased population in the plots treated with the higher amount of PCP is composed of the bacterial cells selected by PCP (i.e. cells highly tolerant to PCP). Figure 6

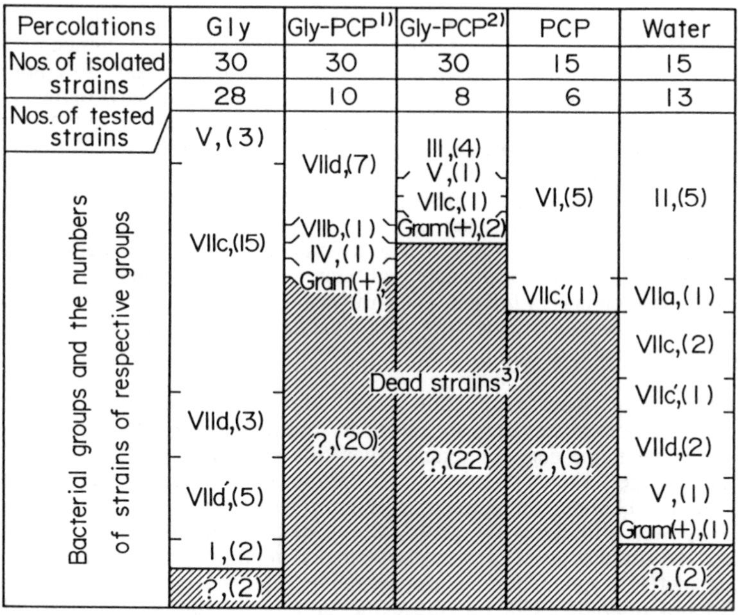

Percolations	Gly	Gly-PCP[1]	Gly-PCP[2]	PCP	Water
Nos. of isolated strains	30	30	30	15	15
Nos. of tested strains	28	10	8	6	13
Bacterial groups and the numbers of strains of respective groups	V,(3)	VIId,(7)	III,(4) V,(1) VIIc,(1)	VI,(5)	II,(5)
	VIIc,(15)	VIIb,(1) IV,(1) Gram(+),(1)	Gram(+),(2)	VIIc',(1)	VIIa,(1)
					VIIc,(2)
			Dead strains[3]		VIIc',(1)
	VIId,(3)	?,(20)	?,(22)	?,(9)	VIId,(2)
	VIId',(5)				V,(1)
	I,(2)				Gram(+),(1)
	?,(2)				?,(2)

Figure 5. A schematic depiction of soil bacterial flora and predominance of respective bacterial groups(5).

Numbers in parentheses after each group indicate the number of strains of the group in both live and dead strains. 1) five day's percolation, 2) fifteen day's percolation, and 3) strains that died after purification.

Genera subsumed in each group: I, *Flavobacterium*; II, *Acinetobacter*; III, *Achromobacter*; IV, not subsumed; V, not subsumed VI, *Pseudomonas*; VII, *Pseudomonas* groups other than that of group VI; VIIa, VIIb, VIIc, VIIc', VIId, and VIId' denote different types of *Pseudomonas* within group VII.

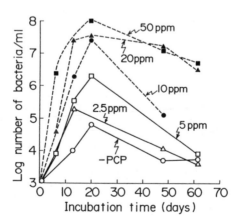

Figure 6. Viable counts of bacteria tolerant to 100 ppm PCP in soil suspensions treated with different concentrations of PCP.

shows the changes in the numbers of bacterial cells tolerant to 100 ppm PCP. The larger the amount of PCP applied, the greater the increase in the numbers of cells, suggesting that the increased proportion of bacterial cells highly tolerant to the pesticide indicates the degree of the impact of the pesticide on the environment. When some microbial nutrients were added together with PCP, the increase in the number of highly tolerant cells was depressed, suggesting that the impact is moderated by the addition of substrates. This coincides also with the pattern of the bacterial flora established in the soils treated with PCP and PCP plus glycine as mentioned in the preceeding section. It is noted that, although the increase is very low, the number of cells tolerant to 100 ppm PCP also increased in the soil without the addition of PCP. Some portion of the tolerant cells may proliferate without the stimulation of pesticide, but the proliferation is very limited. The enrichment of the pesticide-tolerant microbial cells may be adopted as an index suggesting the degree of impact of the pesticide on the soil environment.

PCP-Degrading Bacteria. It is a well known phenomenon that pesticide-degrading microorganisms increase in soil treated repeatedly with a pesticide. This suggests that pesticide-degrading microorganisms are enriched by the addition of the pesticide. Figure 7 shows that there is an optimum PCP concentration for effective proliferation of PCP-degrading microorganisms. Although an increase in the amount of PCP applied stimulates the proliferation, concentrations above 5 to 10 ppm depressed the stimulation. PCP was also degraded in proportion to the proliferation of PCP-degrader(s) in soil. Our previous research revealed that bacteria highly tolerant to PCP did not degrade PCP, but those moderately tolerant to the pesticide did, suggesting that the degrading microorganisms differed from the highly tolerant ones (16). The character of the degrader(s) relates to conditions such as pesticide concentration.

The degradation of PCP was also influenced by the addition of microbial nutrients. PCP-degrading bacteria did not proliferate upon addition of nutrient broth even under optimum PCP concentrations.

The above results suggest that some different processes may operate for proliferating pesticide-degrading microorganisms and pesticide-tolerant ones. Furthermore, the processes are modified by the presence of some microbial nutrients.

Effect of Nutrients on Degradation of PCP in Soil

Adopting the optimum concentration of PCP, the behaviors of, and degradation of PCP by, the PCP-degrading microorganisms were investigated under several nutritional conditions.

Effect of Nutrient Type. Total viable counts increased rapidly after incubation in all experimental plots, and the counts attained were higher in plots treated with the nutrients than plots without nutrients (Figure 8). The nutrients dissipated with time of incubation, indicating that the bacteria proliferated and utilized the added nutrients. The pattern of increase was also very similar to that of the gram-negative microorganisms described above. On the contrary, proliferation of PCP-degrading microorganisms was depressed by the presence of the nutrients (Figure 9). PCP dissipated completely in soil without nutrient addition, indicating that the dissipation was controlled by the proliferating pesticide-degrading microorganisms (Figure 10). PCP also dissipated in plots treated with the nutrients, though the dissipation was incomplete. As the pesticide-degrading microorganisms did not proliferate at all in these plots, chemical degradation might have played a significant role.

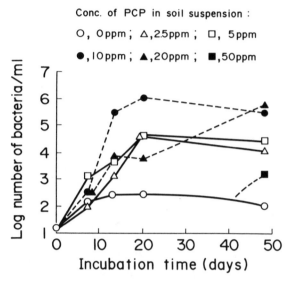

Figure 7. Numbers of PCP-degrading microorganisms in soil suspensions treated with different concentrations of PCP.

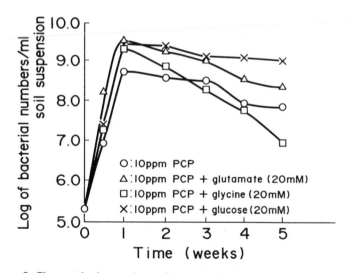

Figure 8. Changes in the numbers of total viable bacteria in soil suspensions containing 10 ppm PCP supplemented with different kinds of nutrients.

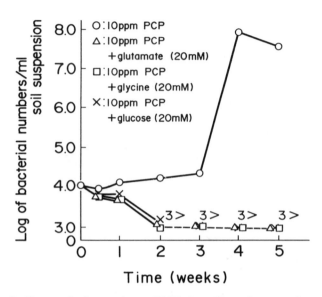

Figure 9. Changes in the numbers of PCP-degrading microorganisms in soil suspensions containing 10 ppm PCP supplemented with different kinds of nutrients.

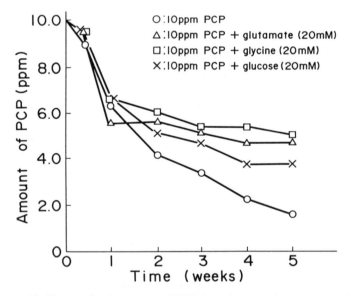

Figure 10. Changes in the amount of PCP in soil suspensions treated with different kinds of nutrients.

Effect of Nutrient Concentration. It is conceivable that the degradation is also influenced by differences in concentration of the nutrients. This view is important for cosidering the degradation of pesticides in natural environments because availability of organic compounds which may be utilized as microbial nutrients varies depending on location. Growth of PCP-degrading microorganisms was retarded by the addition of a large amount of glucose compared with smaller amounts (Figure 11). Initially, total viable counts increased with increasing amounts of glucose added. The delayed increase in the numbers of the PCP-degrading microorganisms in soils amended with glucose began approximately at the time when the rate of CO_2 evolution from the soils became similar to that from the unamended soil. This suggests exhaustion of the added glucose, and also inhibition of growth of pesticide-degrading microorganisms by the presence of microbial nutrients.

Effect of Nutrient Degradability. The effect of differences in nutritional characteristics was compared using glucose and cellulose. The pattern of increase in the number of the PCP-degrading microorganisms was somewhat complex, but a trend was observed. In general, larger populations were observed in the plots without the addition of nutrient or with the addition of cellulose rather than glucose prior to week 3 of incubation (Figure 12). The total viable counts and CO_2 evolution rate were highest in the plot treated with glucose. CO_2 evolution in the plots treated with cellulose, however, was quite similar to that in the plots without nutrients until 4 weeks, when CO_2 evolution markedly increased in the plots treated with cellulose. These results suggest that cellulose was not metabolized soon after the addition, and, therefore was inert in influencing the behavior of pesticide-degrading microorganisms.

Some studies have proposed that the degradation of pesticides in soil was influenced by the organic content of the soil (*17*). The results of the present study coincide with this proposition. However, the present results indicate that the influence of the organic compounds may be related to utilizability, especially to pesticide-degrading microorganisms. Therefore, insight into the nature of the organic compounds is important in deciding their effect on degradation of pesticides in soil. The influence may not be limited to the inhibition of the degradation, but also should be extended to stimulation because cometabolism of pesticides is a well known phenomenon (*18*).

Conclusions

Many studies have been conducted on the modification of soil microflora or the degradation of pesticides for the assessment of pesticide application to soil environments. The present study, however, revealed the soil microflora and pesticide degradation are modified differently by pesticides in the presence of some microbial nutrients than by the addition of only the pesticide. This should be regarded as important to the study of the influence of pesticides on microorganisms in cultivated soil. In many cases the pesticide is appplied to growing plants for protection against pests. As plants exude many kinds of organic materials which may be utilized by microorganisms in rhizosphere soil, consideration of microbial nutrients should be regarded in the study of the relationships between pesticides and soil microorganisms. Further studies should be undertaken to establish a general relationships between the patterns of the microflora and the type of nutrient in the presence of the pesticide.

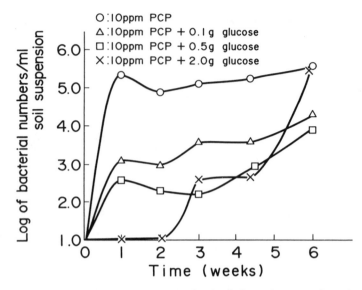

Figure 11. Changes in the numbers of PCP-degrading microorganisms in soil suspensions with different treatments.

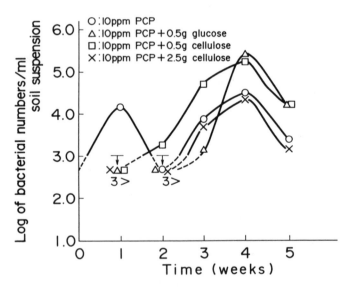

Figure 12. Changes in the numbers of PCP-degrading microorganisms in soil suspensions treated with cellulose or glucose.

Literature Cited

1. Watanabe, I. *Soil Sci. Plant Nutr.* **1973**, *19*, 109.
2. Suzuki, T.J. *Pest.Sci.* **1983**, *4*, 419.
3. Sato, K. *Plant Soil.* **1983**, *75*, 417.
4. Sato, K. *Rep. Inst. Agric. Res. Tohoku Univ.* **1988**, *37*, 15.
5. Sato, K. *J. Gen.Appl.Microbiol.* **1985**, *31*, 197.
6. Sato, K. *Biol. Fertil. Soils.* **1987**, *5*, 1.
7. Russell, E.J. *J. Agric. Sci.* **1913**, *5*, 152.
8. van Schreven, A.D.; Lindenbergh, D.J.; Koridon,A. *Plant Soil.* **1970**, *33*, 513.
9. Tu, C.M. *Canad. J. Microbiol.* **1973**, *19*, 855.
10. Voets, J.P.; Meerschman,P.; Verstraete,W. *Soil Biol. Biochem.* **1974**, *6*, 149.
11. Gunner, H.B.; Zuckerinan,B.M.; Walker,R.W.; Miller, C.W.; Deubert, H.K.; Longley, E.R. *Plant Soil.* **1966**, *25*, 249.
12. Houseworth, L.D.; Tweedy, B.G. *Plant Soil.* **1973**, *38*, 493.
13. Kato, H.; Sato, K.; Furusaka, C. *Bull. Inst. Agric. Res. Tohoku Univ.* **1980**, *32*, 1
14. Sato, K. *Bull. Inst. Agric. Res. Tohoku Univ.* **1971**, *2*, 93.
15. Sato, K. *Canad. J. Microbiol.* **1987**, *33*, 819.
16. Izaki, K.; Takahashi, M.; Sato, Y.; Sasagawa, Y.; Sato, K.; Furusaka, C. *Agric Biol Chem.* **1981**, *45*, 765.
17. Kuwatsuka, S.; Igarashi, M. *Soil Sci. Plant Nutr.* **1975**, *21*, 405.
18. Horvath, R.S. *Bacteriol. Rev.* **1972**, *36*, 146.

RECEIVED April 25, 1994

Chapter 5

Intact Rhizosphere Microbial Communities Used To Study Microbial Biodegradation in Agricultural and Natural Soils

Influence of Soil Organic Matter on Mineralization Kinetics

David B. Knaebel[1] and J. Robie Vestal[2,3]

[1]Department of Bacteriology and Biochemistry, University of Idaho, Moscow, ID 83843
[2]Department of Biological Sciences, University of Cincinnati, Cincinnati, OH 45221-0006

The rhizosphere has been shown to have a higher and more active microbial biomass than surrounding bulk soil. In agricultural and natural soils that harbor plants, this zone may affect the fate of organic chemicals in the soil. In this study, corn and soybeans were grown in an agricultural soil and a woodlot soil; the experimental chambers were designed to contain the soil and roots, but exclude the above-ground plant biomass. Controls contained soils without plants. The biodegradation of several chemicals (anionic, cationic, and nonionic surfactants) was measured. The rhizosphere generally increased initial rates of mineralization, but the total amounts mineralized were not significantly different from controls. This implies that rhizosphere microbial communities metabolize foreign chemicals at greater rates than those in the bulk soil. Since the total amounts mineralized were the same as the controls, however, biodegradation of these chemicals in soils may be limited by other soil - chemical interactions.

The fate of man-made chemicals in surface soils is of unique importance, due to the high potential for these chemicals to interact with humans or other organisms. Soils used for agricultural purposes, industrially-impacted soils, and natural soils are the major categories of surface soils that may be contaminated by a pollutant chemicals. These soils are generally treated with, or exposed to wide ranges and concentrations of chemicals from various sources. In this paper, we address the influence that plants may have on the microbial biodegradation of these chemicals in different types of soils.

Agricultural soils are generally treated with substantial amounts of pesticides and fertilizers which may indirectly affect human health through leaching or transport into aquifer systems. Agricultural soils may also be treated with sewage

[3]Deceased

sludges to increase organic matter content and soil fertility. These sludges may contain considerable amounts of organic contaminants, notably surfactants, as a result of incomplete biodegradation during the sewage treatment process (*1*). Industrially - impacted soils can be contaminated with a wide range of chemicals due to manufacturing, refinement, or other processes. The soils may also be corrupted by spills and unregulated disposal practices. Natural soils, defined as soils not maintained for agricultural, private, or commercial use, can also be impacted by man-made chemicals by accidental or illegal dumping, or by irregular management practices.

It has been found that the fate of chemical contaminants in soils is a function of abiotic and biotic processes, including photo-degradation, leaching, immobilization, and biodegradation. The biodegradation of chemicals in soils usually depends upon the chemical, physical, microbial, and climatic characteristics of the particular soil. In order to understand the microbial metabolism of chemical contaminants, myriad research efforts have studied these processes under wide ranges of conditions. Biodegradation experiments have usually been conducted with only consideration of soil microbiota in the presence (or absence) of the soil matrix. However, to fully address the environmental fate of chemicals in soils that contain plants (i.e., agricultural soils, natural soils), biodegradation experiments must include plants to understand the effects that these organisms might have on the biodegradation process. In agricultural situations, the chemicals will likely interact with the crop plants in the area known as the rhizosphere, which is the area of the soil that surrounds the root tissues. This zone commonly has higher microbial biomass, microbial activity, and higher levels of oxygen and organic carbon than the surrounding bulk soil (*2, 3, 4, 5*). Due to these characteristics, the rhizosphere will likely influence the metabolism of soil contaminants differently than the bulk soil.

The best experimental design for rhizosphere studies should maintain the spatial and temporal association between the soil microbiota and plant root tissues. This is because the removal of soils from the area near the viable plant tissues disrupts the spatial and temporal influences of the roots on the soil chemistry. In addition, even gentle removal of plant tissues from soils likely changes the oxygen content of the soil and results in the fragmentation of fine root tissues and root hairs [for example, small disturbances of sediments prior to sampling have been shown to have effects on the microbial community (*6*)]. An atypical supply of oxygen and carbon would modify the microbial community activity and structure. Furthermore, the loss of the viable plant tissue would cause a cessation in a viable rhizosphere effect, which would diminish the ability to make comparisons to *in situ* processes. Therefore, the influence of the rhizosphere's effect on biodegradation processes is best studied in experiments that maintain the integrity of the rhizosphere during the course of the measurements. Use of ^{14}C-substrates allows biodegradation to be studied by measuring the formation of $^{14}CO_2$, or by collection of ^{14}C-metabolites. This requires a closed microcosm that permits the monitoring of biodegradation products, plant growth, and mass balances. One such microcosm that we developed is shown in Figure 1. This microcosm contains a known amount of soil, a closed atmosphere for the sampling of $^{14}CO_2$, and only a small amount of above-ground plant biomass. This system can easily be modified to permit the monitoring of biodegradation in the presence of the rhizosphere created by a variety of plants. The only constraint on this type of microcosm is that the majority of the plant tissue be outside of the soil atmosphere containment system, such that plant uptake of microbially-generated $^{14}CO_2$ is minimized or prevented.

In agricultural soils, these microcosm chambers should ideally contain plants from the particular crop of interest. For natural soils, the major ground cover, or the plant with the highest community importance should be used, since it likely has the greatest influence on the soil processes. When several plant species are considered to have equivalent importance, then they would ideally be studied in combination chambers, as well as individually. These chambers allow for representative rhizosphere microbial communities to develop that are based on the indigenous soil microbiota (when the seed surfaces are presterilized). Inoculated seeds can also be used to study the effects of exogenous microbes on the development of the rhizosphere and on biodegradation kinetics.

Since plant roots also produce several types of exudates, these should be considered in determining vegetation effects on the fate of the chemicals of interest. These exudates are generally neutral sugars or amino acids or complex carbohydrates (5,7) which provide additional carbon to the microbial community. The exudates contribute to the formation of humic and fulvic acids, which act as sinks for adsorbing other organic compounds (8). Therefore, the primary exudates (or their modified products) may act as sinks into which the chemical contaminants may partition. The fate of these chemicals would then be a function of the degradability of this matrix, as well as a function of the desorption kinetics of the chemical out of the matrix.

Materials and Methods

The materials and methods used in the rhizosphere experiments have been reported elsewhere (9). The details of the experiments that examined the influence of types of soil organic matter on biodegradation are to be published separately (Knaebel et al. *in press*). For clarity's sake, the methods are summarized below.

Soils. Two soils were used: a woodlot soil (Bonnell) and an agricultural soil (Rossmoyne), both collected in southwestern Ohio. The agricultural soil had received annual sludge amendments of approximately 4600 kg ha^{-1} for five years. Samples were collected aseptically, partially dried under a stream of sterile air, and sieved (< 2 mm). Experimental volumes of soil were based upon oven-dried weights (GDW).

^{14}C **Chemicals.** The ^{14}C chemicals were different types of surfactants that may be present in sewage sludges that are added to agricultural soils (Figure 2). [U-^{14}C-ethoxy] C_{12}-linear alcohol ethoxylate (LAE) (specific activity of 5.4 µCi mg^{-1}) was purchased from Amersham (Arlington Hts, Ill.). [U-^{14}C-ring] C_{12}-linear alkylbenzene sulfonate (LAS) [specific activity of 67.8 µCi mg^{-1}] was purchased from New England Nuclear (Wilmington, Del.). [U-^{14}C] sodium stearate (specific activity of 196.8 µCi mg^{-1}) was purchased from Amersham. [1-^{14}C] C_{12}-trimethyl ammonium chloride (TMAC) (specific activity of 95 µCi mg^{-1}) was synthesized at Procter and Gamble, Cincinnati, Oh. [U-^{14}C] acetate (Amersham) was used for the microbial activity measurements and had a specific activity of 57 mCi mMol^{-1}.

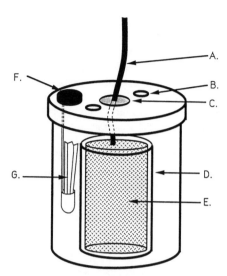

Figure 1. The rhizosphere mineralization chamber used in this study. A, plant stem; B, silicone rubber sealed port; C, hole for plant stem; D, inner chamber containing soil (E); F, stopper in port allowing access to $^{14}CO_2$-capturing wick; G, wick saturated with alkali for capturing $^{14}CO_2$. Controls had no plant, and C was sealed. (Reproduced with permission from ref. 9. Copyright 1992 National Research Council, Canada.)

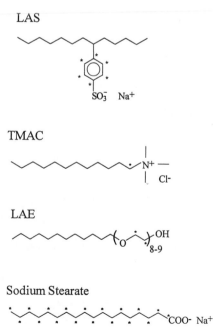

Figure 2. The structural formulas of the chemicals used in this study. The location of the ^{14}C is indicated by *.

Microbiological and Chemical Analysis. Soil microbial activity and biomass were determined prior to, and at the end of, the mineralization experiments. Activity was measured as the rate of ^{14}C-acetate incorporation into microbial lipids (*10, 11*); biomass was measured as $CHCl_3$-extractable lipid phosphate (*12*). Cation exchange capacities of the soils (CEC) were measured by a modification of the method of Bascomb (*13*). Total organic carbon (TOC) was determined gravimetrically following combustion at 550°C for 2 hours. The soil pH was determined with 1:3 soil: distilled deionized water (DDW) slurries.

Mineralization Chambers. A mineralization chamber (Figure 1) was designed that consisted of a glass chamber (50 ml) placed within a larger glass chamber (250 ml). The inner chamber was filled with an amount of the partially air-dried soils that corresponded to 40 GDW. The lid to the larger chamber had a hole for the plant stem to pass through, two ports, and a hole for access to the $^{14}CO_2$ trapping wicks.

Plants. Soybean (*Glycine max*) and corn (*Zea mays*) seeds were purchased from Carolina Biological Supply Company (Burlington NC). Seeds were surface-sterilized (*11*) and aseptically transferred to sterile test tubes containing sterile paper towel fragments saturated with 1.5 ml sterile DDW (SDDW). They were allowed to germinate in the dark until approximately 15 cm tall.

Seedlings were gently transferred to mineralization chambers, such that the leaves and stem passed through the outer lid, and the roots were placed in the soil. The area between the stem and the edge of the hole in the lid was sealed with cotton and high vacuum silicone grease (Dow Corning, Midland MI). The soil was immediately watered with SDDW to bring the soils to 70% gravimetric water holding capacity (WHC). The plant chambers were placed under a combination standard fluorescent - plant light (General Electric) fixture and within a plant growth chamber. The light intensity was maintained at ca. 110 μEinsteins m^{-2} sec^{-1}. The plants were kept on a 14:10 hour light:dark cycle, at room temperature (22-27°C.)

Rhizosphere microbial communities were allowed to develop around the root tissues for 10 days before adding the ^{14}C-surfactants. During the course of the experiments, soil water losses due to transpiration were replaced with SDDW via ports in the lids of the chambers. Water losses were monitored by daily visual inspection of the chambers. Live controls (containing soil only) were treated identically to plant chambers, except that the hole for the stem was sealed with a rubber stopper, and it was not necessary to give additional SDDW to control for transpiration losses. Abiotic soil controls were autoclaved (135°C, 18 psi) for 90 minutes on three successive days prior to ^{14}C-surfactant additions. Also, in place of the SDDW additions, they received a 4.5% (vol/vol) formaldehyde, 0.2 % thimerosol (Sigma) solution to maintain sterility throughout the experiment. Each treatment consisted of 3 to 5 replicates; 2 out of 74 plant replicates had plants which died during the course of the experiments; these were eliminated from analysis.

^{14}C-Surfactant Additions and Mineralization Detection. After 10 days of rhizosphere development, the ^{14}C-surfactants were added to the soils at a concentration of 50 ng g^{-1} soil.. The ^{14}C-surfactant solution (2 ml) was dispersed into the soil with a syringe as uniformly as possible without disturbing the integrity of the root-soil orientation. Microbial mineralization of the surfactants was detected by capturing $^{14}CO_2$ on filter papers saturated with 300 µl 2 N NaOH. Wicks were removed periodically and radioactivity determined. Quench correction was by the external standards method. Mineralization was measured for a period of 3 to 5 weeks. At the end of the mineralization experiments, soils were sampled for percent moisture, microbial activity, and microbial biomass. Soils in live and abiotic controls were mixed with a sterile spatula prior to sampling in order to mimic the disturbance effect of plant root removal.

Soil organic matter-surfactant complexes. Humic acid was purchased from Aldrich Chemical Co, Milwaukee, Wisc. Fulvic acids were extracted from an alpine tundra soil collected from Niwot Ridge, Colo. (*14*) according to standard methods (*15*). Complexes of the humic and fulvic acids with the surfactants were made by mixing sterile aliquots of the organic materials with each surfactant in a slurry of excess aqueous ethanol or methanol. This was placed in sterile flasks on a rotary evaporator and the materials mixed at 40°C. The solvent was slowly evaporated under vacuum as mixing continued. The adsorbed surfactant/organic matter complexes were further dried and stored in a dessicator in sterile containers.

Specific activity determinations of the bound surfactants were made by placing an aliquot in 0.5 ml methanol or ethanol in a 20 ml scintillation vial (Kimble). These were mixed for 5 minutes, and then 15 ml Scintiverse II (Fisher) was added. Water was added to cause the cocktail to gel. Radioactivity determinations were made on a Packard Tri-Carb 2200CA scintillation counter, quench correction was by an efficiency tracing protocol.

Mineralization assays: Mineralization of the complexed surfactants was monitored in serum bottle radiorespirometers as described in Knaebel and Vestal (*11*). Approximately 0.10 g of the adsorbed surfactants was added to 8.0 GDW of each soil. The samples were vortexed for 10 seconds to mix the adsorbed surfactants throughout the soil. SDDW was then added to bring the soils to 70% gravimetric water holding capacity (WHC), and the samples were vortexed again. Non-adsorbed surfactant controls were treated similarly, except that the surfactant was added in an aqueous solution to the soils and then vortexed for 10 seconds. The surfactant was added at a concentration of 50 ng g^{-1} soil. Abiotic controls were prepared by repeated autoclaving and chemical means as previously described (*11*). Mineralization was measured over a period of 2 months.

Data Analysis. Mineralization data were fitted to a first order (*16*) and a mixed (or 3/2) order model (*17*) by a nonlinear regression program [NONLIN; SYSTAT (Systat, Evanston, IL)]. The first order model has the form:

[1] $$P = P_0 (1 - e^{(-k_1 t)}),$$

where P is the percent DPM recovered as $^{14}CO_2$ at time t, P_0 is the asymptotic yield of $^{14}CO_2$ production (%), k_1 is the first order rate constant (day^{-1}), and t is time.

The mixed (or "3/2") order model (with a linear growth term) has the form:

[2] $$P = P_0(1-e^{(-k_1 t - (k_2 t^2)/2)}) + k_0 t,$$

where k_2 represents the linear growth rate and k_0 represents the turnover rate of indigenous carbon in the soil (percent day^{-1}). If no account is made for growth, the growth term is dropped from the model (17); the model becomes:

[3] $$P = P_0(1-e^{(-k_1 t)}) + k_0 t.$$

This form of the 3/2 order model is similar to the first order model, but contains a linear term (k_0) that describes the more slowly-occurring mineralization of microbially-incorporated or microbially-bound ^{14}C. Alternatively, this term may describe the process of mineralization of a compound that is slowly desorbed from the environmental matrix (18).

Models were evaluated for best fit to the data by the F-test method cited in Robinson (19). Statistical analyses were done with the GLM modules of SAS (SAS Institute, Cary, NJ). Differences between mean values were determined with the Ryan-Einot-Gabriel-Welsch (REGW) multiple comparison test (20). Results were considered significant when $p < 0.05$.

Results

The pH of the soils differed greatly, with the Bonnell soil being slightly alkaline (pH=7.9) and the Rossmoyne soil being acidic (pH=4.9). The TOC of the soils also differed greatly, with the Bonnell having a TOC approximately 3-fold higher than the Rossmoyne (10.1 % vs. 2.9%). The CEC of the soils, however, did not differ as much, with the Bonnell soil having a CEC of 27 meq 100g^{-1} and the Rossmoyne soil having a CEC of 19.8 meq 100g^{-1}.

The mineralization chamber was designed to permit sampling of the atmosphere surrounding the soil without substantial interaction from the above ground plant biomass (i.e. CO_2 uptake). The glass construction of the containers permitted visual inspection of the growth of the roots and the gross root morphology, as well as soil moisture qualitative evaluation. This information was used to estimate the amount of water to be added to the chambers, which occurred every 1 to 2 days. Based on total amounts of sterile water added to the soil chambers and the final measured percent soil moisture, the soil moisture averaged approximately 70% WHC. However, this would have been higher immediately following watering and lower just prior to watering. At no time was the soil flooded, such that anaerobiosis-generating conditions were avoided. Light was barred from the chambers during incubation by the lids, as well as by the enclosure that held the chambers.

The presence of the rhizosphere in these soils caused changes in the microbial biomass and activity of these soils. All of the incubation treatments (rhizosphere and soil alone) in the Bonnell soil caused a 2 to 3-fold increase in microbial biomass; only the rhizosphere treatments caused a slight increase in biomass (~ 20 %) in the Rossmoyne soil (Figure 3a). Although the biomass of the control treatment in the Bonnell soil increased during the experiment, the rhizosphere treatments had slightly greater increases. The microbial activity of the Bonnell soil increased in response to the rhizosphere treatments when calculated on a per soil mass basis (Figure 3b). The

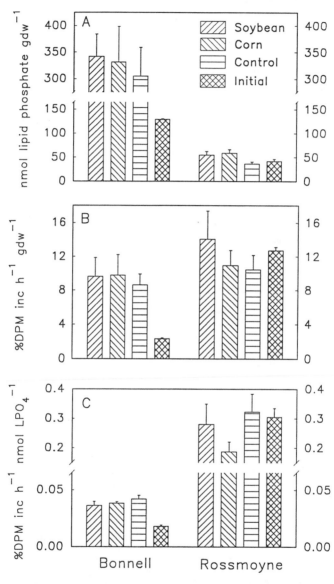

Figure 3. The microbial biomass and activity of the soils before (initial) and after the rhizosphere (soybean and corn) and control mineralization experiments. A., biomass is shown as the mean nmol lipid phosphate per GDW ($\pm$ 1 sd); B., activity measured as the mean percent of $[^{14}C]$acetate incorporated into microbial lipids per hour per GDW ($\pm$ 1 sd); C., activity measured as the mean percent of $[^{14}C]$acetate incorporated into microbial lipids per hour per nmol lipid phosphate [nmol LPO_4] ($\pm$ 1 sd). (Adapted from ref. 9.).

Rossmoyne soil microbial activity increased slightly in the corn treatment as compared to the control treatment, but only the soybean treatment had a significant increase in activity over the control treatment. However, in comparison to the initial microbial activity measurements on this soil, there was a decrease in activity in the corn and control treatments. Due to the design of these experiments, the control treatments more likely reflect the baseline microbial activity of the soil than the initial measurement, due to the initial wetting event and incubation factor.

When the Bonnell microbial activities were calculated on the basis of microbial biomass, the control treatments had slightly greater activities than the soybean and corn treatments (Figure 3c). In the Rossmoyne soil, this calculation resulted in the corn treatment microbiota having a substantially lower activity than the other treatments.

The mineralization data were fitted to the mineralization models, and the 3/2 order model fit significantly better than the first-order model (Table I.)

Table I. Kinetic parameters[a] for the mineralization of the surfactants in the two soils as affected by a plant-induced rhizosphere (soybean or corn) or in soil alone (control)

Soil[b]	Surf[c]	Soybean			Corn			Control		
		P_0	k_1	k_0	P_0	k_1	k_0	P_0	k_1	k_0
Bonn	LAS	32.0	1.16	0.13	32.5	1.02	0.11	37.1	0.43	0.16
		2.7	0.31	0.01	5.5	0.38	0.02	1.6	0.06	0.04
	LAE	41.0	1.51	0.06	40.2	1.20	0.05	45.1	1.31	0.13
		1.2	0.37	0.01	2.0	0.30	0.01	1.6	0.04	0.02
	SA.	27.7	0.30	0.16	23.3	0.38	0.18	26.6	0.15	0.11
		1.2	0.14	0.08	1.3	0.06	0.02	4.5	0.04	0.07
	TMAC	15.6	0.37	0.06	10.4	0.31	0.04	17.4	0.33	0.11
		2.0	0.04	0.01	3.0	0.07	0.01	2.7	0.04	0.02
Ross	LAS	41.9	1.31	0.07	35.3	1.34	0.08	44.9	0.86	0.14
		1.7	0.05	0.02	6.5	0.19	0.01	0.3	0.05	0.01
	LAE	37.7	0.46	0.05	34.6	0.59	0.06	42.8	0.28	0.07
		1.3	0.10	0.03	1.9	0.12	0.02	3.0	0.01	0.01
	SA.	ND[d]			65.6	1.05	0.18	63.7	0.74	0.24
					14.6	0.05	0.01	3.8	0.02	0.02
	TMAC	10.3	0.11	0.0	8.7	0.11	0.0	14.9	0.08	0.02
		1.4	0.01	0.0	0.6	0.01	0.0	0.9	0.01	0.02

SOURCE: reprinted with permission from ref. 9. Copyright 1992.
a. Kinetic parameters are P_0, asymptotic yield; k_1, initial rate of mineralization, k_0 turnover rate of mineralization (see text for explanation) of 3 to 5 replicates. Data are presented as the mean (first line) and standard deviation (second line) for each sample.b. Soil used in this study: Bonn, Bonnell; Ross, Rossmoyne. c. Surfactant mineralized, For abbreviations, see text; SA, Stearate. d. (ND) not determined.

The mineralization of these chemicals in the presence of the rhizosphere was somewhat faster than when there was no rhizosphere present. LAE was mineralized faster in the Rossmoyne soil when it was in the presence of the rhizosphere; in the Bonnell soil, there was no difference in the initial rates of mineralization (k_1) (Figure 4a). The kinetic estimates of the 3/2 order model confirm this observation (Table I). The final amounts mineralized (P_0) were between 34 to 45% conversion to $^{14}CO_2$. LAS, in contrast, was mineralized faster in the rhizosphere treatments in the Bonnell soil. In the Rossmoyne soil, only the soybean showed an apparent increase in the rate of mineralization (Figure 4b). When fitted to the 3/2 order model, however, the estimates of the initial rates were greater in both rhizosphere treatments (Table I). LAS was mineralized to similar extents as LAE (32 to 45 % conversion to $^{14}CO_2$). Sodium stearate was mineralized faster in the presence of the rhizosphere in both soils (Figure 4c, Table I). It was mineralized to similar final extents, however, regardless of treatment. Stearate was mineralized to much greater extents in the Rossmoyne soil than in the Bonnell soil. In contrast to the other surfactants, TMAC was mineralized faster and to greater extents in the bulk soils than in the rhizosphere treatments (Figure 4d). The kinetic estimates (Table I) show no difference in initial rates (k_1) between treatments. The turnover rate component (k_0) of the mineralization of all of the surfactants was consistently lower in the rhizosphere treatments than in the controls (Table I).

The surfactants that had been adsorbed to the different soil constituents were usually mineralized to substantially lesser extents and at slower rates compared to when they were added to the soils in aqueous solutions. LAE was mineralized only after a short lag when bound to either humic or fulvic acids (Figure 5). A similar response was observed for LAS, except that the lag was much longer for the humic-bound LAS, and there was no mineralization of the fulvic-bound LAS (Figure 5). Similar trends were observed for both sodium stearate and TMAC (data not shown).

Discussion

These experiments have shown that there is an effect of the rhizosphere on the mineralization of surfactants in surface soils. When these chemicals were added to the soils at trace amounts (50 µg g^{-1}), the anionic and nonionic surfactants (LAS, stearate, and LAE) were mineralized faster than when added to bulk soils that did not contain a viable rhizosphere community. The cationic TMAC was mineralized similarly regardless of the treatment. The rhizosphere treatments had no significant effect on the overall extent of mineralization of all of the surfactants compared to the bulk soils, which suggests that other phenomena may be affecting the mineralization of these chemicals in these soils. However, other researchers (*21, 22, 23*) have shown greater enhancement of biodegradation in the presence of the rhizosphere. This disparity may have been due to the relatively low levels of chemical added in this study (50 µg g^{-1}) compared to other studies, as well as the nature of the soils used. The low amounts of the chemicals added in this study represent relatively low loading of the chemical, and substantial interactions with the soil organic matter may have occurred. In addition, the intact, viable rhizosphere ecosystems used in these experiments may have caused decreased mineralization of the chemicals due to interactions between the plant, its exudates, the microbiota, and the chemicals (see

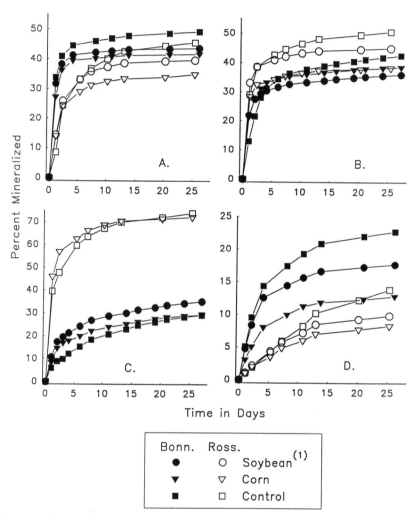

Figure 4. Mineralization of LAE (A), LAS (B), stearate (C), and TMAC (D) as affected by the rhizosphere treatments (soybean and corn) and control treatments in the Bonnell and Rossmoyne soils, n=3-5. Bonn., Bonnell; Ross., Rossmoyne. (1) There was no soybean treatment for the mineralization of stearate. (Adapted from ref. 9.).

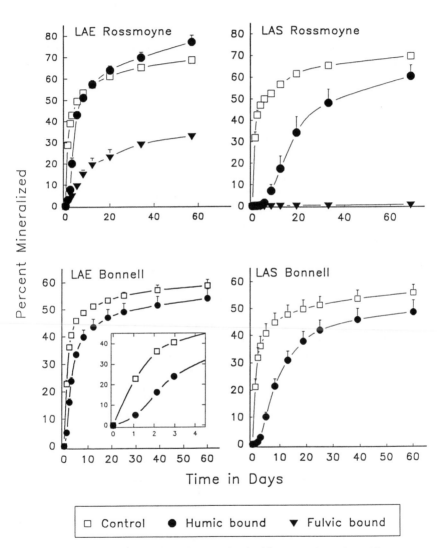

Figure 5. Mineralization of LAE and LAS that had been pre-bound to either humic or fulvic acids in the Rossmoyne and Bonnell soils, n=4. Error bars show 1 sd.

below). In experiments where rhizosphere soil is disrupted and then used for mineralization experiments, the plant influence is removed, and this likely changes the dynamics of the activities of the microbial communities. Further research is planned with these chemicals in soils to address their fate under higher loadings and with a greater viable plant root biomass.

Rhizosphere - Organic Matter Influences. Since one of the documented activities of the plant roots is the release of several organic chemicals into the surrounding soil, there may have been an interaction (either direct or indirect) between this material and the surfactants. In order to address the interactions that organic materials may have on the fate of surfactants (or other contaminants) another set of experiments was conducted. In this experiment, the surfactants were adsorbed to complex organic constituents of soils (humic and fulvic acids) and the mineralization of the bound surfactant was then assessed. In all cases, these bound chemicals had delayed or retarded mineralization, suggesting that interactions between soil organic matter and chemicals of interest can limit their biodegradation (Figure 5). Therefore, it is conceivable that as the plant root tissues released their exudates, some of these interacted with the ^{14}C-surfactants and retarded their access to the microbial community. Alternatively, the exudates may have provided a large labile pool of carbon that was used preferentially for microbial growth and activity. The ^{14}C-surfactants may have been utilized by the microorganisms, but due to the abundance of the labile carbon exudates, the surfactants represented a much lower proportion of available carbon. This may have resulted in the incorporation of the ^{14}C into microbial biomass rather than conversion to ^{14}CO$_2$. These phenomena (the adsorption of the surfactants by the root exudate/organic material, or the dilution of the surfactant carbon by the root carbon) could explain the lower overall amounts of mineralization of the surfactants in the rhizosphere treatments, as well as the reduced turnover rates (k_0) observed in the rhizosphere treatments (Table I).

Although these soil organic matter constituents are not identical to the chemicals exuded from plant roots, they do represent a large component of the ultimate fate of the plant root exudates and of the soil microbial biomass. Therefore, the humic and fulvic acids were chosen to represent the long-lived organic matter pools of soils, and would therefore likely represent pools to which the adsorbed ^{14}C-surfactants would have the longest residence times. Had other organic materials been chosen for the binding matrix (polysaccharides, neutral sugars, or amino acids) the inhibitory effects of the matrix would probably not have been as great. Therefore, the results shown here represent the most severe effects on the biodegradation that would be seen in agricultural or natural soils. The effects and analyses of these and other soil constituents on the mineralization of surfactants in soils will be published elsewhere (Knaebel et al. unpublished results). Other studies are planned that will model the effects of these root exudates on the fate of xenobiotics in viable rhizosphere systems.

Literature Cited.

1. Malz, F.; Korber, H-G. In *Treatment and use of sewage sludge and liquid agricultural waste;* L'Hermite, P. Ed.; Elsevier Science Publ.: New York, NY, **1991**, pp 159-165.
2. Barber, D.A.; Lynch, J.M. *Soil Biol. Biochem.* **1977**, *9*,305-308.

3. Bowen, G.D. In *Contemporary Microbial Ecology;* Ellwood, D.C.; Hedger, J.N.; Latham, M.J.; Lynch, J.M.; Slater, J.H., Eds.; Academic Press, New York, NY. **1980.** pp 283-304.

4. Curl, E.A.; Truelove, B. *The rhizosphere;* Springer-Verlag: New York, NY **1986.**

5. Whipps, J.M.; Lynch, J.M. *Adv. Microb. Ecol.* **1986,** *9,*187-244.

6. Findlay, R.H.; Pollard, P.C.; Moriarty, D.J.W.; White, D.C. *Can. J. Microbiol.* **1985,** *31,*493-498.

7. Martin, J.K. *Soil Biol. Biochem.* **1977,** *9,*1-7.

8. Rovira, A.D; Mcdougall, B.M. In *Soil Biochemistry.* McLaren A.D.; Peterson, G.H. Eds.; Marcel Dekker: New York. **1967.** pp 417-463.

9. Knaebel, D.B.; Vestal, J.R. *Canadian J. Microbiol.* **1992,** *38,*643-653.

10. McKinley, V.L.; FederleT.W.; Vestal, J.R. *Appl. Environ. Microbiol.* **1982,** *43,*129-135.

11. Knaebel, D.B.; Vestal, J.R. *J. Microbiol. Methods* **1988,** *7,*309-317.

12. Findlay, R.H.; King, G.M.; Watling, L. *Appl. Environ. Microbiol.* **1989,** *55,*2888-2893.

13. Bascomb, C.L. *J. Sci. Food Agri.* **1964,** *15,*821-823.

14. Knaebel, D.B.; Federle, T.W.; Vestal. J.R. *Environ. Toxicol. Chem.* **1990,** 9,981-988.

15. Schnitzer, M. In *Methods in soil analysis, part 2, chemical and microbiological properties.* 2nd edition. Page, A.L.; Miller, R.H.; Keeney, D.R., Eds.; American Society of Agronomy, Inc: Madison, WI. **1982.** pp 581-586.

16. Larson, R.J. In *Current perspectives in microbial ecology;* M.J. Klug and C.A. Reddy. Ed.; American Society for Microbiology: Washington D.C., **1984,** pp. 677-686.

17. Brunner, W.; Focht, D.D. *Appl. Environ. Microbiol.* **1984,** *47,*167-172.

18. Sparks, D.L. *Kinetics of soil chemical processes;* Academic press: New York, NY.**1989.**

19. Robinson, J.A. *Adv. Microb. Ecol.***1985,** *8,*61-114.

20. SAS Institute, Inc. . SAS user's guide: statistics, version 5 edition. SAS institute: Cary NC. **1985.** 956 pp.

21. Federle, T.W.; Schwab, B.S. *Appl. Environ. Microbiol.* **1989,** *55,*2092-2094.

22. Hsu, T.-S.; Bartha, R. *Appl. Environ. Microbiol.* **1979,** *37,*36-41.

23. Walton, B.T.; Anderson, T.A. *Appl. Environ. Microbiol.* **1990,** *56,*1012-1016.

RECEIVED March 15, 1994

Chapter 6

Influence of Plant Species on In Situ Rhizosphere Degradation

Jodi R. Shann and J. J. Boyle

Department of Biological Sciences, University of Cincinnati, Cincinnati, OH 45211−0006

The degree of xenobiotic degradation in rhizosphere soil appears related to the species of plant involved. Species of plants are known to vary in their morphology, primary and secondary metabolism, and in their ecological interactions with other organisms. Current investigations examine differences in the rate and quality of root exudates, the adsorption characteristics of root cell walls, and the uptake of xenobiotics by plants. These factors vary with plant species and are correlated to changes in rhizosphere degradation. It is suggested that root exudation may influence degradation by directly altering the microbial community. Cell wall adsorption and plant uptake may limit microbial degradation by controlling the bioavailability of xenobiotics within the rhizosphere ecosystem.

Over the last four decades, an estimated 6 billion tons of hazardous waste has been released in terrestrial ecosystems. In 1991 alone, approximately 3.5 billion pounds of this type of material was released nationally (1). This activity has resulted in widespread contamination of the environment, especially soil systems. Since many organic contaminants are carcinogenic, recalcitrant, and tend to bioaccumulate, their presence in the environment poses a significant hazard to human health. Cleanup, or remediation, of this contamination has become a national and global priority (2).

Conventional soil remediation technologies, such as physical removal of the sediment and washing the soil with solvents, are costly, unpredictable, and frequently insufficient. In recent years, land treatment approaches to remediation that increase natural biodegradation rates, have been investigated. A number of these techniques involve microorganisms cultured and kept in "bioreactors". The contaminated soil must then be brought to the reactor for cleanup. Alternatively, *in situ* remediation has been attempted in various ways, including inoculation with degrading microorganisms. The *in situ* cleanup of contaminated soil has met with varying degrees of success (3).

One limitation to the development of effective *in situ* bioremedial systems is our fundamental lack of understanding concerning the requirements and dynamics of microorganisms in soil ecosystems. In the case of the Exxon Valdez oil spill cleanup, where inoculation with microorganisms seemed to result in soil remediation,

0097−6156/94/0563−0070$08.00/0

increased microbial degradation of the contaminants was observed only when growth-supporting fertilizers were also added. Observations such as this have led to the development of a plethora of commercially available soil "additives" (primarily forms of nutrients) that are advertised as enhancers of biodegradation. In addition, soil aeration and other management techniques are being pursued as another means of supporting microbial activities and therefore increasing biodegradation. These examples, along with numerous others, suggest the growing acknowledgment that bioremedial systems may only become effective and practical after we identify the factors which contribute to the growth and function of microorganisms within their natural environments.

The Rhizosphere

In soil, microorganisms can either be free-living or found associated with other organisms, notably plants. Microorganisms occur with, and are influenced by, these plants in the rooting zone of the soil known as the rhizosphere. Since the rhizosphere is a natural soil habitat that (generally) supports the growth of soil microorganisms, it may prove to be an ecosystem conducive to xenobiotic degradation. Degradation of organic contaminants in the rhizosphere has, to some extent, been investigated in isolated soils and intact plant/microbial systems, and has been inferred from decreases in compound concentrations under vegetated conditions. Since agricultural soils have historically been subjected to chemical inputs, and since crop plants are an integral part of managed ecosystems, many rhizosphere degradation studies have focused on pesticides (4-7). Recent studies, however, have also suggested an enhanced degradation of other xenobiotics (8-11).

The rhizosphere consists of both biotic and abiotic parts. The fate of any xenobiotic introduced into this soil zone is dependent on the potential of the individual components (plant, microorganisms, soil) and on their interactions. Clearly these components are subject to external fluctuations in the physical and chemical environment, but they are also capable of altering their environment. Mechanisms proposed to account for the increased degradation observed in vegetated systems or in rhizosphere soil assume that it is associated with the input of carbon compounds by plant roots. These compounds may then: (1) support greater numbers of organisms (including degraders), (2) provide growth substrate for co-metabolizing organisms or (3) act as structural analogues to xenobiotics and select for degrading populations by enrichment. In pursuit of mechanistic evidence, little attention has been given to the plant factors which ultimately determine what form the plant-derived carbon exudates will take, the type of microbial interaction supported or inhibited by the plant, and the transformation or removal of xenobiotics by the plant itself. If it can be argued that understanding the ecology of degrading microorganisms is critical to the success of *in situ* bioremedial systems, then similar information on the other major biotic component of the rhizosphere, the plant, is equally needed. In the remainder of this paper, we will consider some of the ways in which the species or type of plant may influence the degradation of organic xenobiotics. Our investigations into these areas are discussed.

Plant Factors

The species or general type (monocot, dicot, legume, halophyte, etc.) of plant determines, to a great extent, the parameters that are significant in terms of rhizosphere degradation. For any given species of plant there is a relatively well defined life history (*eg.* annual, perennial), structural morphology - including that of the root (*e.g.* fibrous, tap, deep, shallow), type of microbial associations (*e.g.* endo or ecto-mycorrizal, symbiotic, parasitic), and general/unique lines of secondary metabolism.

The significance of plant type on degradation in rhizosphere soil can be seen in Figure 1. Rhizosphere and non-rhizosphere soils were collected (for complete methods see Boyle, J.J. and Shann, J.R., *J.Environ. Qual.* and *Plant Soil*, in press) in the field from two different soil types and eight different plant species. One half of the species (on each site) were monocots, the remainder dicots. The fresh-collected soils were then used in serum bottle studies where full mineralization of [U-ring ^{14}C]xenobiotic was measured as evolved $^{14}CO_2$. Mineralization of four structurally-related compounds (phenol, 2,4-dichlorophenol [2,4-DCP], 2,4-dichlorophenoxyacetic acid [2,4-D] and 2,4,5-trichlorophenoxyacetic acid [2,4,5-T]) was determined over time.

Although all soils previously associated with plant roots mineralized 2,4-D and 2,4,5-T faster and to a greater extent than non-vegetated soils, those collected from monocot species outperformed those from dicots. No significant difference ($p \leq 0.05$) in mineralization was observed between soils for phenol or 2,4-DCP. This same outcome was also observed in microcosm studies where weak nutrient solutions were flowed over the roots of sand-cultured plants and onto soil columns (Figure 2). Subsamples of the soil were then removed and used in serum bottle studies similar to those described above. The results of those mineralization experiments are given in Figure 3. As in the field-collected soils, the presence of plants in the microcosm increased mineralization. As before, this increase was greatest in the soils receiving exudates from the monocot species.

The above data was generated using soil removed from field-grown or sand-cultured roots and thus represents the residual effect of the plant type; it is not an *in situ* measure of degradation. Addition of the xenobiotic to the soil after removal of the plant, separates the physical presence of the plant from its chemical influence. The plant would, most likely, influence rhizosphere chemistry by production and release of root exudates. A brief review of our current understanding of root exudation and its variation with plant species is discussed below and is followed by some of our investigations into the potentially important physical factors that may also vary with species.

Chemical Plant Factors in the Rhizosphere

Root Exudates. Plant-derived releases into the rhizosphere (rhizodeposition) may be organic or inorganic (13, 14). The release may occur as an active exudation, a passive leaking, the production of mucilage, or with the death and sloughing of root cells. Releases increase under a variety of conditions, particularly forms of abiotic and biotic stress (15). Mechanically stimulated roots release greater amounts of both organic and inorganic material; phenolic exudates from stimulated plants were increased ten-fold over that from unstimulated ones (16, 17). The presence of microorganisms has also been shown to alter plant input of either organics (18) or inorganics (19, 20) into the rhizosphere.

The carbon in root exudates comes from CO_2 photosynthetically fixed in the production of energy-rich carbohydrates. Although reports are often contradictory, anywhere from one to more than forty percent of net photosynthate may be released from roots into the soil (21-24). Organic rhizosphere exudates may take several forms: low molecular weight compounds (*e.g.* simple sugars, amino acids, fatty acids, organic acids, phenolics) or higher-weight polymers such as the polysaccharides and polygalactic acids of mucilage (15, 19).

While living plants exude and leech compounds from their roots, compounds are also released by the decay of dead plant cells. Structural materials such as lignin

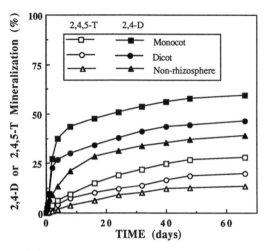

Figure 1. Mineralization in field-collected soils given [1.2 x 10^{-3} µmoles] 2,4-D (closed symbols) or [2.6 x 10^{-3} µmoles] 2,4,5-T (open symbols). Square symbols represent mean values from the pooled monocot species: *Echinochloa crusgalli, Panicum dichotomiflorum, Phleum pratense* and *Setaria viridus*. Round symbols represent mean values from the pooled dicot species: *Trifolium pratense, Erigeron annuus, Solidago juncea* and *Chicorium intybus*. All mean differences are significant at p≤0.05. (Adapted from Boyle, J.J.; Shann, J.R. *J. Environm.Qual.*, in press)

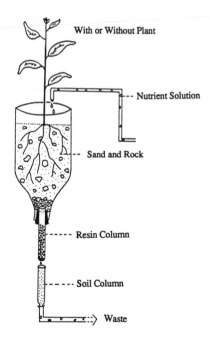

Figure 2. Plant microcosm to remove and/or trap root exudates. Resin columns may or may not contain selective resins. (Adapted from ref. 12)

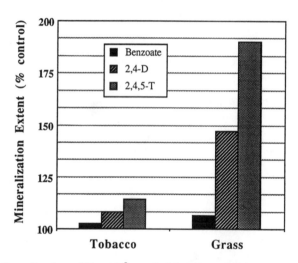

Figure 3. Mineralization of [2 x 10^{-3} μmoles] benzoate, 2,4-D, or 2,4,5-T in soil exposed to root exudates from a dicot (tobacco) or a monocot (grass) plant. Mineralization is expressed as a percentage of the control, *i.e.* soil receiving nutrient flow through an unplanted microcosm. (Haws, K; Shann, J.R., unpublished data)

and cellulose are degraded by physical means and by microorganisms. Cellulose is a linear polymer of glucose and as such is degraded much more rapidly than lignin. Lignin is a huge, random phenolic polymer oxidatively decomposed to stable humic substances with a variety of intermediates and products. Lignin degradation also results in the release of numerous simple phenolics, such as the benzoic and cinnamic acids (25).

Variation With Plant Species. Although the *in situ* function of root exudates has not been fully elucidated, it is clear that (depending on their nature) they can mobilize essential nutrients from the soil, act as antibiotics, and as chemoattractants. Because of their significance to agriculturally important crops, only plant exudates which increase available nutrients, protect against pathogens, or are produced by legumes to signal symbionts and induce infection (26, 27) have been well characterized. The qualitative assessment of less specific root exudates and the quantitative analysis of the exudation process has not been undertaken for a wide range of plant species. Investigations have, however, suggested that exudates vary with plant species; closely related taxa have more similar patterns than do distant ones (28) and responses to environmental conditions appear to be species specific (29). This species variation can also be grouped by plant type; both Fe-deficient dicot and monocot roots accumulate organic acids, but only the dicots have developed a high capacity for proton release into the rhizosphere (30). The significance of plant species or type on exudation is clearest in the highly co-evolved plant-microbial system found in legumes. In any case of co-evolution, the survival of the relationship is dependent on the specific nature of the interaction. In the legume system, the interaction is maintained by chemical cues that must be unique to insure success.

In addition to potential differences in exudates from plant species, structural materials are known to differ between groups such as gymnosperms and angiosperms (31). The difference in the composition of materials such as lignin may result in a species dependent variation in the release of compounds upon decay.

Current Studies. To address the issue of differential exudation by plant species and its consequence on rhizosphere degradation of xenobiotics, we are currently conducting studies in the microcosm shown in Figure 2. In a variation of the experiment presented above, the flow coming out of the sand culture crosses a column containing selective resins prior to the soil column. These resins are used to remove various fractions from the root exudates. The remainder of the exudate is allowed to enter the soil column. In this manner, the influence of specific exudates on serum bottle mineralization of xenobiotics can be quantified and the "active fractions" determined. However, even more significantly, the exudates from any given plant can be concentrated, characterized, and compared to those of similarly grown, but different, species. In separate greenhouse investigations, several species are being grown in the same soil and the same species is being grown in several soils, to separate the plant effect from other soil factors. Resins incorporated into these pots are being used to trap the organic compounds present in the various combinations of soils and species.

Physical Plant Factors in the Rhizosphere

While much attention has been given to the chemical environment of the rhizosphere, the physical one is often ignored in the literature. While many physical factors are relevant in the rhizosphere, this discussion will be limited to the influence the physical presence of the plant has on potential biodegradation. The physical plant presence is considered in terms of how it affects the bioavailability of xenobiotics in the rhizosphere, either by fixation on the roots or uptake by the plant. In addition, it should be noted that the relationship between the physical morphology of the root and rhizosphere degradation becomes important when designing *in situ* bioremedial systems. Higher root to shoot ratios have been correlated to increased release of carbon from the root and to greater numbers of root-microbial associations (32). Also, the volume of soil encountered by a root will vary with its depth and branching pattern. These factors are often modified by the environmental conditions, but are basically determined by the genetic limits of the species involved. The needs of a given remediation may determine the most desirable root system, but a shallow, unbranched one - even if it supported a highly degradative microbial population - may not exploit enough of the subsurface to be effective.

Bioavailability. The bioavailability of a xenobiotic to both the plant and the microbial community has the potential to alter rhizosphere degradation. Adsorbed xenobiotics may not be degradable by microorganisms or may lower the toxicity of the compound and allow for greater microbial activity. Bioavailability is often defined as the amount of a compound that is present in the soil solution or readily exchanged off of system surfaces. In the rhizosphere, adsorption can occur on several such surfaces: the soil, the bacterial cells and, in the case of vegetated soil ecosystems, the roots of plants. In addition, bacterial cells might also adhere to the soil or to roots, and that interaction may decrease or increase their ability to degrade compounds. Rhizosphere microorganisms that are more tightly associated with the plant root (*e.g.* endomycorrhizae or symbiotic nitrogen-fixing bacteria) may "see" more or less of the contaminant based upon their immobility and the transport of the compound to, or past, them.

Bioavailability in the rhizosphere *per se* has received little research attention, but the general role soil adsorption of xenobiotics plays in their microbial degradation has been considered (33-38). Smith *et al*. (38), investigated the effect of soil adsorption on quinoline (an N-heterocyclic) hydrocarbon degradation by bacterial isolates. Quinoline was not degraded when in the sorbed state. Ogram *et al*. (36) modeled the degradation of 2,4-D by microorganisms if it was adsorbed onto soil and the ability of those microorganisms to degrade free compound if the microbial cells were sorbed. Their data indicated that the fraction of 2,4-D adsorbed onto soil was completely protected from degradation by both sorbed and solution bacteria, as measured by $^{14}CO_2$ evolution from the radiolabeled compound. On the other hand, adsorption of the bacteria did not seem to affect their ability to degrade 2,4-D. However, in a separate study, the addition of microbial biomass to subsurface soil appeared to reduce the adsorption of quinoline and naphthalene. This reduction in xenobiotic adsorption was attributed to a biomass alteration of: (1) the soil surface or (2) access to this surface (37). Binding studies with specific soil materials have also indicated the inaccessibility of sorbed xenobiotics to microbial populations. When 2,4-DCP was either surface-sorbed or incorporated into humic acid, mineralization was inhibited (34). While these studies indicate that adsorption of organic xenobiotics to soil or soil components does make them less available, the implications for long-term *in situ* rhizosphere degradation are not yet well defined. As previously stated, the presence of the root in rhizosphere systems offers another surface on which xenobiotics may be adsorbed and/or transformed. The most likely site for this to occur in a root is on the cell wall.

Cell Walls. The pores in cell walls have diameters in the range of 35-40 angstroms, a size which probably does not pose a substantial physical barrier to most molecules. The cell wall is, however, a source of fixed, non-diffusible charges. Like the surfaces found on soil materials, the primarily negatively charged sites on cell walls may adsorb xenobiotics in the rhizosphere and make them less bioavailable for degradation.

Much is now known about plant cell wall composition and structure (39-41). The wall constituents believed responsible for a majority of the fixed charge are the acidic pectic polysaccharides (galacturonic acids); their apparent pKa values are in the same general range as those of isolated cell walls (as determined by titration) (42). Quantitative differences in these pectin constituents are known to occur between, and within, monocots and dicots (43, 44).

While the plant cell wall does possess some anionic-exchange properties, it has been investigated primarily as a cation-exchanger (45-53). Most of the investigations of cell wall adsorption have focused on inorganic ions, but may offer some insight to the potential for the adsorption of organic xenobiotics. Several of these studies (45, 47, 51) have indicated that the wall is a selective exchanger. In addition to experimental approaches, theoretical models have been used to describe the ionic behavior of cell walls (47, 49, 54, 55). Most models treat the cell wall as a typical Donnan system, although mass-action expressions have also been employed (54). Allen and Jarrell (55) suggest that two basic types of models have been utilized: one which assumes a charged surface on which ions condense till a specific density is reached, and one which incorporates both a Stern layer for selectively adsorbed ions and a diffuse double layer for nonspecific (or electrostatic) adsorption. Allen and Jarrell (55) successfully applied the constant capacitance model to wall adsorption by maize and soybean. Described as a high ionic strength limiting case of the Stern, this model was appropriately used as a predictive tool for the strongly adsorbed Cu ion. Van Cutsem and Gillet (51) suggest the wall contains both high-affinity (inner sphere complexed) and low-affinity (outer sphere, electrostatic complexed) sites for Cu^{2+}. Unlike Cu, the behavior of purely outer sphere complexed ions would not be quantitatively predicted by a model such as the constant capacitance. Ions that form both outer and inner sphere complexes would be better represented by a Stern-like model where selective and electrostatic forces would be accounted for.

The effect of cell walls on 2,4-D mineralization in soil is shown in Figure 4. The walls were isolated (following ref. 54) from 2-day-old seedlings of wheat (*Triticum aestivum* L. cv. Caldwell) and incorporated into serum bottles containing 5 g of field-collected rhizosphere soil. Mineralization of [U-ring^{14}C] 2,4-D was then monitored over time. All soils and treatment solutions were adjusted to pH 7.4, where a majority of the 2,4-D would be in the anionic form. The addition of cell walls to the soil significantly decreased mineralization of 2,4-D in a manner proportional to the amount incorporated. This decrease persisted with subsequent spikes of 2,4-D and appears to represent a removal of some portion of this herbicide from the bioavailable pool. Studies are currently being conducted to determine the adsorption of 2,4-D to sterile preparations of both the isolated cell walls and the soil, and to walls isolated from 2,4-D-exposed wheat seedlings. The cultivars of wheat being used have been shown to vary in the titratable charge characteristics of their root cell wall. Across these cultivars, higher wall charges correlate to greater resistance to mononuclear and polynuclear Al toxicity; presumably a response associated with the fixation of Al on the walls, which then decreases plant Al uptake (56). Again, if wall characteristics differ between taxonomic or functional plant groups, the potential impact of root adsorption on rhizosphere degradation may be dependent on the plant species - or even cultivar - involved.

Plant Uptake. While the adsorption of a xenobiotic to the cell wall of a root may decrease its immediate bioavailability to the remainder of the rhizosphere, there remains the possibility of exchange back into the soil solution. Uptake of the xenobiotic by the plant, however, represents a more permanent removal of the compound from the rhizosphere. Plants have the potential to take up xenobiotics, translocate them, conjugate them, and/or transform them by oxidative, reductive or hydrolytic reactions. Uptake of organic xenobiotics has been reviewed elsewhere (57-

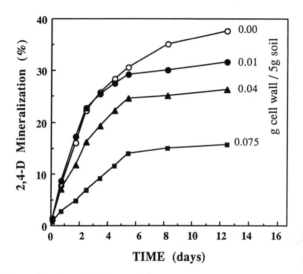

Figure 4. Mineralization of [1.81 x 10^{-4} µmoles] 2,4-D in 5 g of soil amended with 0, 0.01, 0.04, or 0.075 g (dry weight) of wheat cell wall.

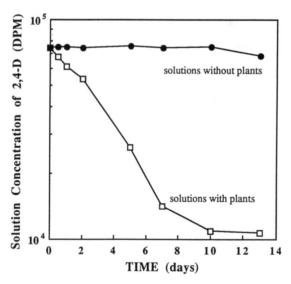

Figure 5. Concentration of 2,4-D in hydroponic solutions removed from tubs containing plants or from those without plants. All tubs initially contained 75,000 DPM [3 x 10^{-4} µmoles] 2,4-D. Solution concentrations were calculated for pooled tubs with-plants or without-plants, regardless of the presence or absence of microorganisms. (Alexander, E.S.; Shann, J.R.. unpublished data).

59). Although the chemical nature of the xenobiotic determines to a large extent its fate within the plant, there is variation in the potential of different species to respond; this variation is the basis of selective herbicidal action. One could generalize and say that herbicides are designed for their activity towards certain taxonomic groups, *e.g.* "broad-leaved" (dicots) or "woody" (also dicots), or their inactivity towards a desirable crop species, *e.g.* the triazines (used on corn).

The significance of plant removal, via uptake, on rhizosphere mineralization was observed in studies conducted in our lab. The original purpose of these experiments was to determine if the presence of the symbiotic state in the plant-microbial association, *Bradyrhizobium japonicum* and soybean, altered rhizosphere mineralization of phenol or 2,4-D. Nodulated/non-nodulated wild-type soybean, and nodulating/non-nodulating mutant soybeans were hydroponically cultured with, or without the *Bradyrhizobium*. The *Bradyrhizobium* was also inoculated into the same systems without any plants present. Evolved $^{14}CO_2$, along with solution and plant tissue radioactivity were measured. Although the mineralization of the phenol and 2,4-D was significantly increased whenever the plants were inoculated with the bacteria, the symbiotic state did not enhance or diminish this effect. More interesting in terms of rhizosphere dynamics, hydroponic tubs containing plants had much lower concentrations of phenol and 2,4-D than did those without them (Figure 5). When tissues were analyzed, over 80% of the added radiolabel was found in the plant. This finding supports the general assumption that the plant will, to some extent, determine how much of the xenobiotic is available for degradation by rhizosphere microorganisms.

Summary

Rhizosphere degradation offers great promise for *in situ* bioremediation. Realization of this promise, however, relies on a fusion of the plant, microbial, and soil disciplines. Here, we have briefly noted one aspect significant to the development of *in situ* systems, the species of plant (or plants) utilized. Degradation and bioavailability of xenobiotics in soil systems may vary with characteristics determined by plant species or type. The studies and literature presented only hint at the potential impact that species differences may have on those factors most significant to the use of rhizosphere degradation as a bioremedial tool.

Literature Cited

(1) Lean, G.; Hinrichsen, D; Markham, A. *Atlas of the Environment*; Prentice Hall: New York, NY, 1990.
(2) Bokairo, A.A. *National Geographic* **1985**, *167,* pp 319-321.
(3) Thomas, J.M.; Ward, C.H.; Raymond, R.L.; Wilson, J.T.; Loehr, R.C. In *Encyclopedia of Microbiology;* J. Ledergerg, Ed.; Academic Press: New York, NY, 1992, Vol. 1; pp 369-385.
(4) Sandmann, E.R.I.C.; Loos, M.A. *Chemosphere* **1984**, *13,* pp 1073-1084.
(5) Hsu, T.S.; Bartha, R. *Appl. Environ. Microbiol.* **1979**, *37,* pp 1225-1228.
(6) Reddy, B.R.; Sethunathan, N. *Appl. Environ. Microbiol.* **1983**, *45,* pp 826-829.
(7) Lappin, H.M.; Greaves, M.P.; Slater, J.H. *Appl. Environ. Microbiol.* **1985**, *49,* pp 429-433.
(8) Federly, T.W.; Schwab, B.S. *Appl. Environ. Microbiol.* **1989**, *55,* pp 2092-2094.
(9) Walton, B.T; Anderson, T.A. *Appl. Environ. Microbiol.* **1990**, *56,* pp 1012-1016.

(10) April, W ; Sims, R.C. *Chemosphere* **1990**, *20*, pp 253-265.
(11) Knaebel, D.; Vestal, J.R. *Can. J. Microbiol.* **1992**, *38*, pp 643-653.
(12) Tang, C; Young, C. *Plant Physiol.* **1982**, *69*, pp 155-160.
(13) Rovira, A.D.; Harris, J.R. *Plant and Soil.* **1961** *14*, pp 199-214.
(14) Rovirra, A.D.; Foster, R.C.; Martin, J.K. In *The Soil-Root Interface;* Harley, J.L.; Russell, R.S., Eds.; Academic Press: New York, NY, 1979.
(15) Marschner, H. *Mineral Nutrition of Higher Plants;* Academic Press: New York, NY, 1986.
(16) Schonwitz, R.; Zielgler, H. *Z. Pflanzenphysiol.* **1982**, *107*, pp 7-14.
(17) Lameta D'Arcy, A. *Plant Soil* **1982**, *68*, pp 399-403.
(18) Prikryl, Z.; Vancura, V. *Plant Soil* **1980**, *57*, pp 69-83.
(19) Curl, E.A.; Truelove, B. *The Rhizosphere*, Springer-Verlag: New York, NY, 1986.
(20) Barber, D.A.; Martin, J.K. *New Phytol.* **1976**, *76*, pp 69-80.
(21) Gregory, P.J.; Atwell, B.J. *Plant Soil* **1991** *136*, pp 205-213.
(22) Martin, J.K.; Kemp, J.R. *Soil Biol. and Biochem.* **1986** *18*, pp*103-107*.
(23) Martin, J.K. In *Soil Organic Matter Dynamics and Soil Productivity - a Review;* J.H. Cooley, Ed.; International Association for Ecology: Athens, GA, 1987; pp 17-23.
(24) Lynch, J.M. In *Ecology of Microbial Communities*; Fletcher, M.; Gray, T.R.G.; Jones, T.G., Eds.; Cambridge University Press: Cambridge, MA, 1987.
(25) Flaig, W. *Geochim. Cosmochim. Acta.* **1964**, *28*, pp 1523-1535.
(26) Phillips, D.A. *Recent Adv. Phytochem.* **1992**, *26*, pp 201-231.
(27) Leon-Barrios, M; Dakora, F.D; Joseph, C.M.; Phillips, D.A. *Appl. and Environ. Micro.* **1993**, *59*, pp 636-639.
(28) Rovira, A.D. *Bot. Rev.* **1969**, *35*, pp. 35-57.
(29) Bachmann, G; Kinzel, H. *Soil Biol. Biochem.* **1992**, *6*, pp 543-552.
(30) Romheld, V; Marschner, H. **1983** *Plant Physiol.* **71**, pp 949-954.
(31) Lewis, D.H. *New Phytol.,* **1980**, *84*, pp 209-229.
(32) Gardner, W.K.; Parbery, D.G.; Barber, D.A. *Plant Soil* **1982**, *68*, pp 19-32.
(33) Scow, K.M.; Hutson, J. *Soil Sci. Soc. Am. J.* **1992**, *56*, pp 119-127.
(34) Dec, J.; Shuttleworth, K.L.; Bollag, J.M. *J. Environ. Qual.* **1990**, *19*, pp 546-551.
(35) Haider, K.M.; Martin, J.P. *Soil Biol. Biochem.* **20**, pp 425-429.
(36) Ogram, A.V.; Jessup, R.E.; Ou, L.T.; Rao, P.S.C. *Appl. Environ. Microbiol.* **1985**, *49*, pp 582-587.
(37) Bellin, C.A.; Rao, P.S.C. *Appl. Environ. Microbiol.* **1993**, *59*, pp 1813-1820.
(38) Smith, S.C.; Ainsworth, C.C.; Traina, S.J; Hicks, R.J. *Soil Sci. Soc. Am. J.* **1992**, *56*, pp 737-746.
(39) Albersheim, P. In *Plant Biochemistry*; Bonner, J.; Varner, J.E., Eds.; Academic Press: New York, NY, 1976.
(40) Fry, S.C. *J. Exp. Bot.* **1989**, *40*, pp 1-11.
(41) Ishii, T; Thomas, J.; Darvill, A.; Albersheim, P. *Plant Physiol.* **1989**, *89*, pp 421-428.
(42) Morvan, C.; Demarty, M.; Thellier, M. *Plant Physiol.* **1979**, *63*, pp 1117-1122.
(43) Jarvis, M.C.; Forsyth, W.; Duncan, H.J. *Plant Physiol.* **1988**, *88*, pp 309-314.
(44) Mort, A.J.; Grover, P.B. *Plant Physiol.* **1988**, *86*, pp 638-641.
(45) Clarkson, D.T. *Plant Soil* **1967**, *61*, pp 731-736.

(46) Dainty, J.; Hope, A.B.; Denby, C. *Aust. J. Biol. Sci.* **1978**, *12*, pp 395-411.
(47) Demarty, M.; Morvan, C.; Thellier, M. *Plant Physiol.* **1978**, *62*, pp 477-481.
(48) Haynes, R.J. *Bot. Rev.* **1980**, *46*, pp 75-99.
(49) Sentenac, H; Grignon, C. *Plant Physiol.* **1981**, *68*, pp 415-419.
(50) Ritchey, R.J.; Larkum, A.W.D. *J. Exp. Bot.* **1982**, *33*, pp 125-139.
(51) Van Cutsem, P.; Gillet, C. *J. Exp. Bot.* **1982**, *33*, pp 847-853.
(52) Demarty, M.; Morvan, C.; Thellier, M. *Plant Cell & Environ.* **1984**, *7*, pp 441-448.
(53) Irwin, P.L.; Sevilla, M.D.; Shieh, J.J. *Biochimica et Biophysica Acta* **1984**, *805*, pp 186-190.
(54) Bush, D.S.; McColl, J.G. *Plant Physiol.* **1987**, *85*, pp 247-260.
(55) Allen, D.L.; Jarrell, W.M. *Plant Physiol.* **1989**, *89*, pp 823-832.
(56) Allen, D.L; Shann, J.R.; Bertsch, P.M; *Plant Soil* **1990**, *25*, pp 345-349.
(57) Ryan, J.A.; Bell, R.M.; Davidson, J.M.; O'Connor, G.A. *Chemosphere* **1988**, *17*, pp 2299-2323.
(58) Paterson, S.; Mackay, D.; Tam, D.; Shiu, W.Y. *Chemosphere* **1990**, *19*, pp 297-331.
(59) Walton, B.T.; Hoylman, A.M. *Electric Power Research Institutee Interim Report* **1992**, *EPRI TR-101651* project 2879-10.

RECEIVED February 15, 1994

Chapter 7

Rhizosphere Microbial Communities as a Plant Defense Against Toxic Substances in Soils

Barbara T. Walton[1], Anne M. Hoylman[2], Mary M. Perez[3],
Todd A. Anderson[4], Theodore R. Johnson[5], Elizabeth A. Guthrie[6],
and Russell F. Christman[6]

[1]Environmental Sciences Division, Oak Ridge National Laboratory,
Oak Ridge, TN 37831–6038
[2]Graduate Program in Ecology, University of Tennessee,
Knoxville, TN 36996–1191
[3]Environmental Engineering Department, Texas A&M University,
College Station, TX 77843
[4]Pesticide Toxicology Laboratory, Iowa State University, Ames, IA 50011
[5]Department of Biology, St. Olaf College, Northfield, MN 55057
[6]Department of Environmental Sciences and Engineering,
University of North Carolina, Chapel Hill, NC 27599–7410

Vegetation often grows and proliferates in the presence of relatively high concentrations of organic compounds in soils. Experimental data from studies of plant-microbe interactions implicate the rhizosphere microbial community as an important exogenous line of defense for plants against potentially harmful chemicals in soils. Experimental data on the fate of numerous industrial and agricultural chemicals in the rhizospheres of several plant species are consistent with the hypothesis proposed herein that the metabolic activity of the rhizosphere microbial community may protect plants from toxicants in soil. Rhizosphere microbial communities are shown to enhance the associations of three ^{14}C-PAHs with the combined humic and fulvic acid fraction of soil. Such an association could also reduce the bioavailability and potential phytotoxicity of the PAHs. Recommendations are formulated for conducting bioremediation studies of plant-microbe-toxicant interactions in the laboratory. These recommendations are a logical consequence of the hypothesis presented herein that the metabolic detoxication capabilities of the rhizosphere microbial community can protect plants against potentially toxic compounds present in soils.

The chapters in this book provide excellent testimony to the considerable value of well-focused experiments designed to understand the dynamics of plant-microbe-toxicant interactions in the rhizosphere. Numerous environmental factors may influence chemical fate in the rhizosphere, and our understanding of the precise role of each variable may determine our success in using vegetation to reclaim chemically contaminated surface soils.

0097–6156/94/0563–0082$08.00/0
© 1994 American Chemical Society

Formulation of a central theme for plant-microbe-toxicant interactions may also prove to be extremely useful for bioremediation, in that such a theme could provide a paradigm to guide hypothesis testing and stimulate thinking on ways to optimize degradation of organic contaminants in the rhizosphere. In this chapter, we propose a generalized toxicological response of plant life to toxic substances in the rhizosphere, whereby the metabolic detoxication capability of the microbial community protects plants by reducing the potential toxicity of organic compounds near the roots. We envision this interaction to be highly dynamic and mutually beneficial to survival of both the host plant and the rhizosphere microbial community.

As a prelude to further elaboration of this idea, we present a brief overview of biochemical defenses in other biota.

Biological Defenses Against Toxicants

The discipline of toxicology is based on a fundamental precept that all substances are poisonous to life if the dose is sufficiently great (*1*). A consequence of this universal potential of any chemical to be a poison is that the unchecked entry of organic compounds into cells can produce toxicity. Toxicologists assert that all life forms have evolved protective mechanisms against toxicants in the environment, moreover, major taxonomic groups appear to use common strategies for protection. Defenses against potential toxicants may be in the form of physical barriers to toxicant entry and physiological or biochemical detoxication, such as sequestration, secretion, excretion, or biotransformation of the toxicant. Some generalizations about these defenses are reviewed.

Animals. The cytochrome P-450 mixed function oxidase (MFO) system can be viewed as a biochemical defense used by multicellular animals, such as mammals, fish, birds, and insects, to detoxicate hazardous organic compounds (*2, 3*). Many of the latter compounds are lipophilic. The MFO enzyme system utilizes a cytochrome P-450 and NADPH cytochrome P-450 reductase to reduce an atom of molecular oxygen to water while the other atom is incorporated into the substrate (*4*). The consequence of this reaction is a more polar molecule that is either excreted directly or further conjugated to endogenous polar chemicals such as glucose, glucuronic acid, sulfate, phosphate, or amino acids. The conjugate is then excreted (Fig. 1). Because the overall process of enzymatic transformation usually reduces the biological activity of the substrate and promotes elimination, biotransformation is usually a detoxication process (*3, 5*). The MFO system is highly inducible upon first challenge with a xenobiotic, thus, energy is conserved in that enzyme synthesis occurs only when the need for detoxication arises.

Bacteria. Aerobic bacteria can also effectively oxidize organic xenobiotics. An important difference from toxicant transformations in multicellular animals is that dioxygenase enzymes are used by bacteria to introduce two atoms of oxygen into aromatic substrates (*6*). This makes bacteria capable of deriving energy from xenobiotics (Fig. 2); thus, under aerobic conditions lipophilic compounds are more likely to be metabolized completely to CO_2 and H_2O, rather than to be excreted as a metabolite of the parent compound. General information about detoxication of organic compounds by rhizosphere microbiota is summarized elsewhere in this volume (*7*).

Plants. Plants have several biochemical and physiological defenses against toxic substances. In addition, plants exclude many non-nutritive organic substances at the root surface (8). Once an organic toxicant enters the root, it may be metabolized or immobilized through a number of processes, including, compartmentalization into storage vacuoles, formation of insoluble salts, complexation with plant constituents, or binding of the compound to structural polymers (9, 10).

Enzymatic oxidation, reduction, and hydrolysis reactions generally reduce toxicity of the substrate. These reactions are mediated primarily through mixed function oxidases, peroxidases, nitroreductases, and hydrolases (9). Subsequent conjugation reactions with water-soluble cell constituents, such as glutathione, amino acids, glucose, or other thiols and sugars (9, 11, 12) typically increase the polarity, water solubility, and molecular weight of the substrate and reduce phytotoxicity (9). The metabolite may then be translocated within the plant and subsequently immobilized. The overall result is reduced potential for toxicity of a xenobiotic to the plant (11).

The metabolism of aromatic and heterocyclic herbicides by higher plants is generally limited to removal of ring substituents and incorporation of the ring structure into plant polymers as a bound residue (9, 13). Ring cleavage and subsequent oxidation

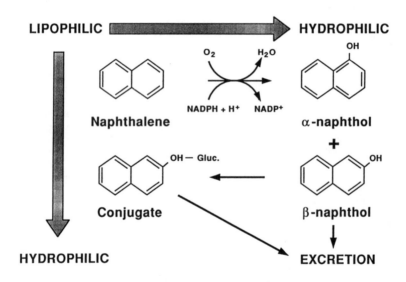

Figure 1. The cytochrome P-450 mixed function oxidase system facilitates excretion of lipophilic organic substrates in terrestrial and aquatic animals by producing metabolites that are more water soluble than the parent compound. Common metabolites of naphthalene are shown (after 2).

of the compound to CO_2, although common in microorganisms, does not appear to be a significant route of transformation by higher plants (*14*).

In summary, plant cells are afforded protection against lipophilic, organic substances that enter plant tissue primarily by compartmentalization and enzymatic transformations. In contrast to multicellular animals in which transformation of the substrate facilitates excretion, compartmentalization and immobilization tend to be the ultimate fate of a toxicant in plants.

Plant-Microbe-Toxicant Interactions in the Rhizosphere

Recent reviews summarize plant-microbe-toxicant interactions in the rhizosphere (*7, 15, 16*). These reviews provide numerous examples of microbial degradation of hazardous substances in the rhizospheres of a variety of plant species. Further evidence of microbial degradation of xenobiotics is provided in other chapters of this volume. We propose that these individual cases of microbial detoxication in the rhizosphere be viewed collectively as evidence for a generalized response of plant life to chemical stress encountered in soil. We speculate that the full range of defenses available to plant life against toxicants may extend beyond the physical barriers and internal responses that have been described. These plant defenses may be augmented by metabolic detoxication capabilities of the rhizosphere microbial community.

The external protection afforded by the rhizosphere microbial community can be envisioned to have evolved to the mutual benefit of both microorganisms and plants. That is, microorganisms benefit from nutritional enrichments in the form of plant exudates, sloughed root cells, dead root hairs, litterfall, and other means by which photosynthate is transferred from the plant to soil (*17*). The plant, in turn, benefits from the metabolic detoxication of potentially toxic organic compounds in soil accomplished by the rhizosphere microbial community. In this scenario, plant expenditures of energy for chemical detoxication are conserved because photosynthate in the form of root exudates maintain the microbial community under normal, unstressed conditions. We speculate that when a chemical stress is present in soil, a plant may respond by increasing or changing exudation (carbon allocation) to the rhizosphere (Fig. 3). As a result, the microbial community, in turn, increases the transformation (and detoxication) rate of the toxicant. This microbial response could be an increase in microbial numbers, an increase in synthesis of detoxication enzymes, or a change in the relative abundance of those microbial strains in the rhizosphere that can degrade the toxicant. Thus, the plant would gain protection by inducing the exogenous metabolic capabilities of the rhizosphere microbial community. Such stimulation can be viewed as analogous to induction of the endogenous MFO system of multicellular animals. In this way, the plant would expend energy to boost or induce enzyme synthesis when the plant is stressed by toxicants in the surrounding soil; at other times, the microbial community would exist at a maintenance level. Thus, we invoke positive feedback, for which there are numerous examples in natural ecological systems (*18*) as the regulatory system for toxicant-plant-microbe interactions.

Evidence for Coevolution of Plant-Microbe Defenses

Proof that plants have a common defense strategy against chemical toxicants whereby the versatile metabolic capability of the rhizosphere microbial community protects the plant, is most likely to come from the weight of experimental evidence consistent with this premise. No singular experiment can either prove or disprove the relative importance of rhizosphere microbiota to detoxication and plant protection. Our formulation of this hypothesis emerged from years of work on plant-microbe-toxicant interactions at Oak Ridge National Laboratory. These studies include experiments on several plant species from a number of field sites using several chemical contaminants, as well as examination

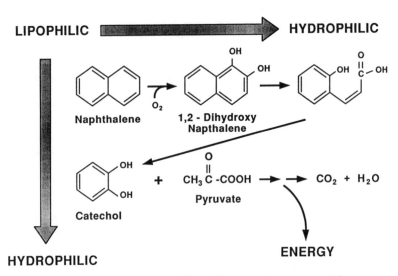

Figure 2. Biochemical transformations of lipophilic aromatic compounds by procaryotes proceed through dihydroxy metabolites of the parent compound. Ring fissions lead to production of metabolites that can be completely mineralized to yield energy, as is shown for naphthalene (after 6).

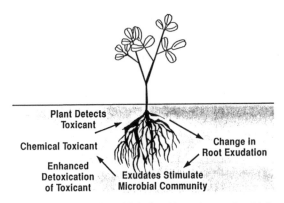

Figure 3. Hypothetical mechanism by which the rhizosphere microbial community may be influenced by the host plant to promote detoxication of an organic substance in soil. By this scenario, a chemical toxicant in the soil would be detected by the plant and the plant would respond with a change in either the quality or quantity of root exudates. This change in exudation would evoke an increase in numbers of rhizosphere microorganisms or an increase in the relative abundance of those strains best able to metabolize the toxicant. The enhanced detoxication accomplished by the microbial community could be the result of mineralization of the substrate, biotransformation to produce a less toxic compound, or stabilization of the toxicant by polymerization reactions, such as humification. This proposed pathway would operate as a positive feedback loop until the concentration of the toxicant in soil was sufficiently reduced that the plant returned to a normal pattern of root exudation.

of the published literature.   The results are reported in a number of theses, dissertations, technical reports, submitted manuscripts, and manuscripts in preparation; however, no single document sets forth the essence of our general idea of plant-microbe-toxicant interactions.

We have formulated observations that would be expected, if the rhizosphere microbial community does protect its host plant from potentially toxic compounds in soil. These observations provide minimal criteria for evaluating whether experimental data are consistent with the protective role of microbe-toxicant interactions in the rhizosphere. The minimal criteria are as follows:

- Mineralization of xenobiotics would occur faster and to a greater extent in the rhizosphere than in comparable bulk soil,
- Many different plant species would show enhanced degradation of xenobiotics in the rhizosphere, that is, the phenomenon would be widespread in the plant kingdom,
- Rhizosphere microbial communities would change quantitatively and qualitatively following exposure to even low concentrations of toxicants,
- Plants would expend energy to support their rhizosphere microbial communities,
- Once plant life is removed from soil, the rate of microbial degradation of xenobiotics in that soil would diminish.

These criteria are statements of what might be expected from plant-microbe-toxicant studies, if rhizosphere microorganisms did coevolve with plant life and a mutualistic symbiosis exists between them. Evidence consistent with the first two criteria comes from the published scientific literature, including many new examples of toxicant degradation in the rhizosphere described in other chapters of this book. Additional evidence comes from studies by Hoylman (*19*), who observed that rhizosphere microbial communities of *Melilotus alba* increased following exposure to 250 µg/g phenanthrene in soil when compared with *M. alba* rhizosphere soil in the absence of phenanthrene. Importantly, Hoylman also found that *M. alba* shifts carbon allocation from aboveground tissue to roots when exposed to phenanthrene in the soil. This carbon shift was measured as a reallocation of ^{14}C-photosynthate to roots and an increase in $^{14}CO_2$ efflux from soil when plants were exposed to phenanthrene as compared with control plants grown in the absence of phenanthrene. These increases in microbial numbers and carbon allocation to the rhizosphere are consistent with our prediction that plants would expend energy to augment a microbial community that protected it from toxicants in soil.

In other experiments using *M. alba*, soils were analyzed for incorporation of ^{14}C into fulvic and humic fractions of soil after treatment of vegetated and nonvegetated soils with four ^{14}C-PAHs. These studies were designed to measure uptake and translocation of ^{14}C-naphthalene, ^{14}C-phenanthrene, ^{14}C-fluoranthene, and ^{14}C-pyrene by *M. alba* over a 5-day exposure. Mass balance of the ^{14}C in these studies revealed that more than 86% of the ^{14}C remained in the soil at the conclusion of the exposure period (*19*). Additional analyses of the soils reveal that ^{14}C from three of the four PAHs become associated with fulvic and humic fractions in the rhizosphere. Data on the association of ^{14}C with fulvic and humic fractions, presented below, indicate that the rhizosphere microbial community may accelerate humification of some aromatic compounds. Association of PAHs with the combined humic and fulvic fraction of soil could also be expected to reduce the potential toxicity of hazardous substances to the host plant.

Humification of PAHs in the Rhizosphere

A composite soil (typic Udipsamments, mixed, mesic) was collected from two depths (0-18 and 25-28 cm) near a coal tar disposal pit in the northeastern United States. The total organic carbon of the composite was $0.93 \pm 0.04\%$ and the pH was 6.98. Particle size distribution was sand, 94.5%; silt, 3.3%; and clay, 2.2%. White sweetclover, *Melilotus alba*, was selected for uptake studies due to its presence at the site, morphological characteristics, and ease of study under laboratory conditions. *Melilotus alba* seedlings were germinated for uptake studies to obtain plants of similar ages and sizes. Prior to exposure to individual ^{14}C-PAHs, *M. alba* plants of approximately the same age (2 months) and size were transferred to 250-mL flasks to acclimate for a minimum of 5 days in a walk-in environmental chamber. Thus, for the vegetated treatments, ^{14}C-PAHs were applied to soil containing a plant with an established root mass. Otherwise, vegetated, nonvegetated, and sterile controls were treated the same.

Four ^{14}C-PAHs, [1-^{14}C]naphthalene (specific activity 318 x 10^{10} Bq per kg); [9-^{14}C]phenanthrene (specific activity 270 x 10^{10} Bq per kg, radiochemical purity >98%, Sigma Chemical Co., St. Louis, MO); [4,5,9,10-^{14}C]pyrene (specific activity 101 x 10^{11} Bq per kg, radiochemical purity >98%; Sigma Chemical Co., St. Louis, MO), and [3-^{14}C]fluoranthene (specific activity 101 x 10^{11} Bq per kg, radiochemical purity >97%, Sigma Chemical Co., St. Louis, MO), were applied individually to soil (60-75 g). The total concentration of PAH applied to the soil was 250 μg per g soil for each PAH, with the exception of phenanthrene, which was applied at a concentration of 500 μg per g soil. ^{14}C-PAHs were dissolved in a minimum volume of toluene and applied to soil in vegetated (soil and plant), nonvegetated (soil only, no plant), and sterile (soil autoclaved for 1 h over 2 consecutive days, no plant) treatments. Nonradiolabeled PAHs were purchased from the following vendors: Aldrich Chemical Co., Milwaukee, WI (fluoranthene); J.T. Baker Inc., Phillipsburg, NJ (naphthalene); and Sigma Chemical Co., St. Louis, MO (phenanthrene and pyrene). All organic solvents were reagent grade and >99% pure.

Soil-plant-flask units were maintained in glass exposure chambers housed in a walk-in environmental chamber. The chamber was programmed for a 14-h photoperiod; diurnal temperatures in the chamber were 32/22 °C. Ambient CO_2 level was 370 μL per L. Glass exposure chambers were kept under negative pressure with exiting air passing through a charcoal filter to trap volatile ^{14}C-compounds released to the atmosphere from the aboveground foliage.

At the conclusion of the 5-day treatment period, plants were removed for analysis (*19*) and soil samples were sequentially extracted with *n*-hexane, chloroform, and 0.5 M NaOH under N_2 (*20*). Approximately 30 g soil and 80 mL of *n*-hexane were added to 110 mL sample bottles and agitated on a rotator/shaker (Thermolyne, Dubuque, IA) at 300 rpm for one hour. Samples were then centrifuged (International Equipment Co., Needham Hts., MA) at 2000 rpm for 10 minutes. Following centrifugation, 200 μL aliquots of the supernatant were dissolved in Aquasol (New England Nuclear, Boston, MA) scintillation cocktail and counted using liquid scintillation spectroscopy (Packard Model 2000CA). The extraction procedure was performed three times with hexane, followed by three times with chloroform. After each extraction, ^{14}C was counted.

After the third chloroform extraction, each soil sample was air dried to a constant weight. Fifty mL of 0.5 N NaOH was added to the dry soil; the unit was then purged with N_2, closed, and agitated at 200-250 rpm for 24 hours (*21*). Samples were centrifuged for 10 minutes at 2,200 rpm to separate the dark-colored supernatant comprising humic and fulvic acids and water-soluble degradation products from the soil residuum. Carbon-14 determinations of this combined fraction were made using liquid scintillation spectroscopy. After acidification to pH 2, water-soluble PAH degradation products (22) were extracted from the combined fraction using hexane. Carbon-14 determinations were made on the extract using liquid scintillation spectroscopy.

Results and Discussion. The percentage of the total extractable ^{14}C associated with the combined humic and fulvic acid (HA/FA) fraction (including water-soluble ^{14}C-metabolites retained in this fraction) was greater in vegetated treatments for all PAHs except pyrene. Pyrene showed no statistically significant influence of vegetation (Table 1). The pH 2 hexane extractions showed that the majority of the ^{14}C in this combined fraction was unextractable and was presumably associated with the combined HA/FA fraction of the soil organic matter. The pH 2 hexane extract was not analyzed for specific water-soluble acids and phenolic acids of the type found by Guerin and Green (22); however, the lack of appreciable ^{14}C extraction after acidification to pH 2 indicates that the ^{14}C in this fraction is not likely to be the result of an increase in such acid metabolites of the PAHs.

The ^{14}C in the combined HA/FA fraction of the sterile soil treatments indicates incorporation of ^{14}C from each PAH under sterile conditions, that is, in the absence of an active microbial population. Only in the case of ^{14}C-naphthalene does a pronounced abiotic association occur (Table 1), which may be related to the smaller size of naphthalene. Comparison of the ^{14}C in the HA/FA fraction of the nonvegetated soil with that of the matched sterile soil provides an indication of the contribution of the microbial community to the fate of ^{14}C-PAHs in the rhizosphere. The fluoranthene, phenanthrene, and naphthalene data are consistent with the hypothesis that microbial processes enhance the association of ^{14}C with the HA/FA fraction in the rhizosphere of *M. alba* during a relatively short (five day) incubation period.

Table 1. ^{14}C-distribution in soil organic matter after a 5-day exposure period in sterile, nonvegetated, and vegetated soil with *Melilotus alba*

[14]C-Compound	Soil Treatment	Fulvic/Humic Acid Fractions (% Total ^{14}C)[A]
Pyrene	Sterile Soil	5.1 ± 2.6
	Nonvegetated	4.4 ± 0.4
	Vegetated	4.3 ± 0.8
Fluoranthene	Sterile Soil	1.4 ± 0.2
	Nonvegetated	2.7 ± 0.5
	Vegetated	5.8 ± 0.2
Phenanthrene	Sterile Soil	1.8 ± 0.3
	Nonvegetated	7.1 ± 4.5
	Vegetated	30.6 ± 13.8
Naphthalene	Sterile Soil	15.6 ± 4.5
	Nonvegetated	14.6 ± 5.9
	Vegetated	40.8 ± 7.0

[A] Data are the mean % ± one standard deviation of the ^{14}C extracted from soils treated with individual ^{14}C-PAHs (three replicates). Percent ^{14}C associated with chloroform and hexane extractions make up the remainder of the extractable radioactivity.

These preliminary findings suggest that the rhizosphere microbial community may influence the fate of selected PAHs in soils, and that the association of these PAHs with the HA/FA fraction of soils occurs at a faster rate in the rhizosphere as a result of microbially mediated processes. However, these preliminary data do not prove that ^{14}C-substrate was chemically or physically incorporated into those macromolecules implied by the operationally defined terms "humic" and "fulvic" acids. Soil organic matter contains a wide range of water-soluble to water-insoluble macromolecular material, most of which is isolated by the standard soil science procedures we employed using NaOH extraction preceded by chloroform/hexane extraction. It is possible, for example, that the rhizosphere microbial community partially oxidized the substrate to more polar water-soluble products which were present in the base extract, without any formal association with humic or fulvic materials. However, the pH 2 hexane extract counts indicated that these products, if present, were in relatively small amounts. Subsequent work should determine the distribution of ^{14}C in high performance size exclusion chromatography profiles of these macromolecules and/or the alteration of their ^{13}C nuclear magnetic resonance spectra to confirm the humification hypothesis.

Incorporations of PAHs into the NaOH-extractable soil fraction would be expected to decrease the biological availability of PAHs in soils and, hence, reduce the potential toxicity of the test compounds to the host plant. Moreover, these experimental findings, when viewed from the perspective of plant-microbe-toxicant interactions, suggest yet another mechanism by which the microbial community may protect the plant from toxicants in soils, that is, by an accelerated humification rate in the rhizosphere.

Coevolution of Plant-Microbe-Toxicant Interactions

Plant symbiotic associations with microbiota are well recognized (23, 24). The hypothesis that mutualism exists between rhizosphere microbiota and host plants as a defense against potential toxicants in soils invites speculation on the selective pressures that would favor coevolution of this symbiosis. It is interesting to speculate that the association may have evolved in soils where naturally occurring, lipophilic organic compounds such as PAHs reach high concentrations. For example, forest and grass fires could provide significant natural sources for PAHs (25) such that survival of individual plants may be impeded after incidents of high PAH loading. Survival advantage would accrue for those plant species able to become established in PAH-laden soils. The release of allelopathic chemicals in the root zone is another example of a naturally occurring chemical stress that would favor rhizosphere microbial communities able to detoxicate organic substrates.

Soil microorganisms may increase plant survival in environments where concentrations of naturally occurring organic compounds constitute a chemical stress to plants germinating at these sites. Moreover, in recent times, microorganisms associated with roots may enhance plant survival in the presence of anthropogenic compounds, such as herbicides, insecticides, surfactants, chlorinated solvents, and chemical wastes.

Implications of Plant-Microbe-Toxicant Interactions for Bioremediation

The idea that rhizosphere microbial communities are beneficial to plant survival is well recognized in many areas of plant science and merits consideration in plant ecotoxicology. The idea that plants may benefit from metabolic detoxication capabilities of their rhizosphere microbial communities provides an interesting paradigm for examining experimental data on plant-microbe-toxicant interactions in soils. If the rhizosphere microbial community of a host plant can influence the fate and bioavailability of toxicants in soils, then adjustments to the design of laboratory studies to evaluate bioremediation potential in the field may be in order.

A number of specific implications of plant-microbe-toxicant interactions for laboratory bioremediation studies become apparent. If the rhizosphere microbial community is influenced both by the plant and the toxicant in soil, then *laboratory studies for bioremediation purposes should be conducted using soils, plants, and rhizosphere microbial communities collected from the actual field site* where preexposure and preselection of toxicant-degrading microbial communities will have occurred.

Studies involving *terrestrial plant species should be conducted using soil* as the rooting medium to permit development of a rhizosphere microbial community. *Hydroponic growth media should be avoided* for bioremediation studies, because rhizosphere microbial communities characteristic of terrestrial plants are more likely to proliferate in soil than in water.

Experimental conditions should be fully described, including plant species, soil characterization, and exposure history. *Controls are needed to distinguish biological from nonbiological losses of test compounds*, and *mass balance determinations of test compounds are essential*. Moreover, based on findings reported herein, *mineralization should not be the only endpoint for bioremediation studies–microbial incorporation of test compounds into humic and fulvic organic fractions of soils should also be considered*, but a reliable method is needed to do this.

Conclusion

The concept that plants may be protected from toxicants in soils by the metabolic detoxication capabilities of their rhizosphere microbial communities provides a fascinating area for biological, chemical, and toxicological research that may have important implications for the use of vegetation *in situ* to decontaminate soils containing hazardous waste chemicals.

Acknowledgments

The authors gratefully acknowledge helpful contributions to this work by Prof. F. K. Pfaender, University of North Carolina, Chapel Hill; Drs. D. L. DeAngelis, S. Wullschleger, S. E. Herbes and C. W. Gehrs, Oak Ridge National Laboratory (ORNL); Dr. Philip Albro, National Institute of Environmental Health Sciences, Research Triangle Park, NC, and Dr. V. J. Homer, Oak Ridge Institute for Science and Education, Oak Ridge, TN. Research sponsored in part by the Electric Power Research Institute, Inc., Research Project 2879-10 to ORNL and the U.S. Department of Energy through the HBCU/MI ET/WM Consortium's EMCOM Scholarship managed by Associated Western Universities. ORNL is managed by Martin Marietta Energy Systems, Inc., under contract DE-ACO5-84OR21400 with the U. S. Department of Energy. Publication No. 4236 of the Environmental Sciences Division, ORNL.

Literature Cited

1. Gallo, M. A.; Doull, J. In *Casarett and Doull's Toxicology: The Basic Science of Poisons*, 4th ed.; Amdur, M. O., Doull, J., Klaassen, C. D., Eds.; Pergamon Press: New York, NY. **1991**, pp. 3-11.
2. Wilkinson, C. F. In *The Scientific Basis of Toxicity Assessment*; Witschi, H., Ed.; Elsevier/North-Holland Biomedical Press: Amsterdam, **1980**, pp. 251-268.
3. Guengerich, F. P. *Am. Sci.* **1993**, *81*, 440-447.
4. Lu, F. C. *Basic Toxicology: Fundamentals, Target Organs, and Risk Assessment*, 2nd ed.; Hemisphere Publishing Corporation: New York, NY. **1991**.

5. Sipes, I. G.; Gandolfi, A. J. In *Casarett and Doull's Toxicology: The Basic Science of Poisons*; 4th ed.; Amdur, M. O., Doull, J., Klaassen, C. D., Eds.; Pergamon Press: New York, NY. **1991**, pp. 88-126.
6. Rochkind, M. L.; Blackburn, J. W.; Sayler, G. S. In *Microbial Decomposition of Chlorinated Aromatic Compounds*. EPA/600/2-86/090. U. S. Environmental Protection Agency: Cincinnati, OH. **1986**, pp. 45-67.
7. Walton, B. T.; Guthrie, E. A.; Hoylman, A. M. In *Microbial Degradation of Organic Chemicals in the Rhizosphere*; Anderson, T. A., Coats, J. R., Eds.; ACS Symposium Series; American Chemical Society: Washington, D.C. **1994**.
8. Walton, B. T.; Edwards, N. T. In *Land Treatment: A Hazardous Waste Management Alternative*; Loehr, R. C.; Malina, J. F., Jr., Eds.; University of Texas: Austin, TX. **1986**, pp. 73-86.
9. Hatzios, K. K.; Penner, D. *Metabolism of Herbicides in Higher Plants*; Burgess Publishing Company: Minneapolis, MN. **1982**, pp. 15-74.
10. Sandermann, H., Jr. *Trends Biochem. Sci.* **1992**, *17*, 82-84.
11. Shimabukuro, R. H. In *Weed Physiology, Vol. II: Herbicide Physiology*; Duke, S. O., Ed.; CRC Press: Boca Raton, Fl, **1985**.
12. Lamoureux, G. L.; Frear, D. S. In *Xenobiotic Metabolism: In Vitro Methods*; G. D. Paulson; Frear, D. S.; Marks, E. P., Eds.; American Chemical Society: Washington, D. C., **1979**, pp. 25-36.
13. Ashton, F. M.; Crafts, A. S. *Mode of Action of Herbicides*; John Wiley & Sons: New York, NY. **1981**, 525 p.
14. Hatzios, K. K. In *Environmental Chemistry of Herbicides, Vol. II.*; Grover, R.; Cessna, A. J., Eds.; CRC Press: Boca Raton, Fl, **1991**, pp. 141-187.
15. Walton, B. T.; Anderson, T. A. *Curr. Opin. Biotechnol.* **1992**, *3*, 267-270.
16. Anderson, T. A.; Guthrie, E. A.; Walton, B. T. *Environ. Sci. Technol.* **1993**, *13*, 2630-2636.
17. Curl, E. A.; Truelove, B. *The Rhizosphere*; Springer-Verlag: Berlin, **1986**, 288 p.
18. DeAngelis, D. L.; Post, W. M.; Travis, C. C. *Biomathematics: Positive Feedback in Natural Systems, Vol. 15*, Springer-Verlag: Berlin, **1986**, 290 p.
19. Hoylman, A. M. Ph.D. Dissertation. Fate of polycyclic aromatic hydrocarbons in plant-soil systems: Responses of plants to a chemical stress in the root zone. University of Tennessee, Knoxville, TN, **1993**, 119 p.
20. Schnitzer, M.; Schuppli, P. *Soil Science Society of America Journal*, **1989**, *53*, 1418-1424.
21. Schnitzer, M.; Schuppli, P. *Canadian Journal of Soil Science*, **1989**, *69*, 253-262.
22. Guerin, W. F.; Jones, G. E. *Appl. Environ. Microbiol.*, **1988**, *54*, 929-936.
23. Ahmadjian, V.; Paracer, S. In *Symbiosis: An Introduction to Biological Associations*; University Press of New England: Hanover and London, **1986**, 212 p.
24. Margulis, L.; Fester, R. In *Symbiosis as a Source of Evolutionary Innovation*; MIT Press: Cambridge, MA, **1991**, 454 p.
25. Edwards, N. T. *J. Environ. Qual.* **1983**, *12*, 427-441.

RECEIVED May 19, 1994

Chapter 8

Potential Use of Mycorrhizal Fungi as Bioremediation Agents

Paula K. Donnelly and John S. Fletcher

Department of Botany and Microbiology, University of Oklahoma, Norman, OK 73019

Currently the most common practice of cleaning contaminated soil is to remove the soil from the site and transfer it to a permanent storage cell, or to an incinerator for combustion of organics. The use of microorganisms as bioremediation agents is gaining popularity based primarily on degradation studies conducted under laboratory conditions. Unfortunately these microorganisms are often not active degraders when moved from the laboratory to the field, presumably because they cannot compete with the native organisms. We are proposing the use of ectomycorrhizal fungi as bioremediation agents with special attention focused on both the fungus and the host plant. The host plant will give the fungus a selective advantage for surviving at a contaminated site. Several of these fungi are known to metabolize various chlorinated aromatic compounds, such as 2,4-dichlorophenoxyacetic acid (2,4-D), atrazine, and polychlorinated biphenyls (PCBs). This group of fungi may play an important role in the bioremediation of hazardous compounds in soil. This chapter summarizes the work to date using mycorrhizal fungi in bioremediation studies.

The rhizosphere is the zone around young and active roots exhibiting increased microbial populations and activity. Root systems continuously release metabolic substrates, especially sugars and amino acids, by both secretion and root turnover (death). Without the release of organic substrates by roots, many microorganisms would not survive in the soil ecosystem (1). Garrett (1) refers to the plant root system as "the most important inhabitants of the soil, because ... they provide the energy that drives the whole of the soil ecosystem." This carbohydrate loss from the plant is not as detrimental to the plant as it may first appear, because many soil microorganisms foster plant growth. The superior uptake properties of some microorganisms in the rhizosphere facilitate the movement of soil minerals into plants that would otherwise be unavailable. Thus the exudates released from the roots stimulate microbial growth which in turn fosters plant growth (1).

Of the fungi that occur in the rhizosphere, there are two important groups that infect the roots of plants. One group is comprised of the pathogenic root-infecting

0097–6156/94/0563–0093$08.00/0

fungi which can cause serious economic losses and widespread damage to many plant species worldwide. The second group of root-infecting fungi, the mycorrhizal fungi, occur in and on the roots of the host plant in a mutualistic, symbiotic relationship. In general, the host plant provides the fungi with a soluble carbon source and the fungi improve the acquisition by the host plant of water and nutrients especially when the water and nutrients are in short supply. In this relationship, both partners benefit from the association (1-4).

Studies with mycorrhizal fungi have focused on their ability to provide the host plant with nutrients in a nutrient-deficient soil (1-5) and have ignored the full role they may play in the soil ecosystem. For example, although laboratory studies have shown that several mycorrhizal fungi have enzymatic properties similar to saprophytic fungi (6-8), it is unclear whether or not these enzymes are active under field conditions. It is possible that although mycorrhizal fungi draw most of their nutritional needs from the host plant, they may retain the ability to degrade a large assortment of compounds available in the soil including organic pollutants. The prospect of such fungi existing in nature has important implications for bioremediation because mycorrhizal fungi introduced in combination with their host plant at a contaminated site may survive and have sustained degradative properties, features often not characteristic of introduced bacteria.

Classification and Growth of Mycorrhizal Fungi

A large number of fungi have been classified as mycorrhizal with a wide variety of host plants (9-12). Some of these fungi have a narrow host range while others have a broad range of host plants (11). Several studies have shown that with higher green plants, the presence of this mycorrhizal association is more common than its absence (4). It should also be mentioned that an individual plant may have more than one mycorrhizal fungal species associated with it at any given time or over a period of time (13).

Traditionally the mycorrhizal fungi were classified into two groups according to the position of the hyphae in relation to the root epidermis of the host plant. Ectotrophic mycorrhizae do not penetrate the host cells but form an intercellular hyphal network and enclose the root in a dense sheath. The endotrophic mycorrhizae penetrate the host cells extensively, but have only a loose network of hyphae on the root surface. Within these two major groups, there were subgroups which contained the ericaceous, orchid, and ectendotrophic mycorrhizae (14). Mycorrhizal fungi are now usually classified as belonging to one of four major types: the ectomycorrhizae, the vesicular-arbuscular mycorrhizae, the ericaceous mycorrhizae, and the orchid mycorrhizae (1,14). The ectomycorrhizal fungi have been successfully grown in axenic culture; however, axenic cultures of vesicular-arbuscular mycorrhizal fungi do not exist (14).

Metabolism of Mycorrhizal Fungi

Catabolic Enzymes. Several species of ectomycorrhizal fungi have been shown to produce the enzymes necessary to degrade complex aromatic compounds in the soil (6,15,16). Bae and Barton (6) reported the following hydrolytic enzymes to be present in *Cenococcum graniforme*: protease, esterase, a-D-galactopyranosidase, b-D-galactopyranosidase, a-D-mannopyranosidase, b-D-xylopyranosidase, a-D-glucopyranosidase, b-D-glucopyranosidase, and alkaline phosphatase. *Hebeloma crustuliniforme* was shown to produce an extracellular acid proteinase (16). Proteinase activity has also now been measured in other ericoid and ectomycorrhizal fungi (17). Amylase, xylanase, mannase, cellulase, and polyphenol oxidase have been found in various ectomycorrhizal fungi (15). Although Hutchison (18) did not find the presence of cellulase, ligninase, or pectinase in selected ectomycorrhizal fungi, several other enzymes were detected. Various ectomycorrhizal fungi produce lipase, amylase, gelatinase, and urease (*18*). Leake and Read (*19*) showed the production of a free acid protease by an ericoid mycorrhizal fungus. Significantly increased levels of protease, phosphatase, peroxidase, xylanase, cellulase, and laminarinase have been measured in the mat structures formed by certain ectomycorrhizal fungi as compared to non-mat soils (*7,8*). The presence of these enzymes indicates that mycorrhizal fungi may not be completely dependent on the host plant for its carbon supply. By producing these enzymes, mycorrhizal fungi have the potential ability to utilize other carbon sources found in the soil ecosystem and may play an important role in nutrient cycling. Whether this actually occurs needs to be studied. However, all indications show that this group of fungi are possibly capable of degrading a wide range of compounds.

Lignin Degradation. In recent years the nonmycorrhizal, ligninolytic white-rot fungus, *Phanerochaete chrysosporium*, has been shown to be efficient at degrading numerous other aromatic compounds (*20-24*). Eaton (*21*) showed that *P. chrysosporium* was able to mineralize PCBs, which are usually highly resistant to microbial degradation. He noted the similarities in the pattern of degradation of this compound with that of lignin, thus supporting the theory that the ligninolytic system was involved (*21*). Haider and Martin (*22*) tested the ability of *P. chrysosporium* to degrade several xenobiotic compounds. This fungus was able to degrade several chlorinated aromatic compounds, and the optimum conditions for degradation, such as low N concentration, high aeration, shaking, and the presence of a cosubstrate, were similar to the optimum conditions for lignin degradation (*22*). However, the optimum conditions required for degradation are not possible under field conditions. Aust (*20*) showed that the ligninolytic system of *P. chrysosporium* was active during the degradation of other complex aromatic pollutants. It appears that the ligninolytic system is nonspecific; the enzymes are expressed regardless of whether the compound is present or not (*20*). These characteristics make *P. chrysosporium*, as well as any other microorganism with a similar nonspecific enzymatic system, good potential bioremediation agents.

Recent studies have indicated the ability of both ericoid and ectomycorrhizal fungi to mineralize lignin (*5,25-28*). Vesicular-arbuscular mycorrhizal fungi have not been tested for lignin degradation because they cannot be grown in pure culture. Trojanowski et al. (*28*) showed certain ectomycorrhizal fungi were capable of degrading various plant cell wall components *in vitro*. Lignin, holocellulose, and lignocellulose were used as substrates and were degraded (*28*). When lignin was supplied as the sole carbon source, both ectomycorrhizal and ericoid mycorrhizal fungi were capable of utilizing the lignin *in vitro* (*25,27*). *Hysterangium setchellii* is an ectomycorrhizal fungus that forms extensive hyphal mats at the soil-litter interface with the host plant Douglas fir (*29*). These mats colonize up to 27% of the forest floor and account for 45 - 55% of the total soil biomass (*30*). Recent evidence indicates that ectomycorrhizal fungal mats may provide an environment that can increase the rate of organic matter decomposition (*31,32*). *Hysterangium setchellii* mat soils were compared to adjacent non-mat soils for their ability to degrade lignin. This was tested seasonally over a one year period. In all seasons, the lignin degradation was higher in the mat soils than in the non-mat soils (*26*).

The combination of recent studies indicate that many mycorrhizal fungi are capable of degrading complex aromatic compounds present in the soil. Many ectomycorrhizal fungi apparently are not completely dependent on their host plant for their carbon supply and are capable of utilizing other available carbon sources to supplement the carbon supply from the host plant.

Environmental Pollutants

Heavy Metals. An important benefit of mycorrhizal infection has been observed in sites with heavy metal contamination, such as mine spoils, and in sites containing high levels of other toxic materials (5,33). Mine spoil sites often have high levels of zinc, manganese, and cadmium. Plants with mycorrhizal fungi appear to be partially protected against the toxicity of the heavy metals. It is believed that the mycorrhizal fungi present in these contaminated sites have been selected over time with greater resistance to these metals. The heavy metals are believed to be bound by carboxyl groups in the hemicelluloses of the interfacial matrices between the host cells and the fungus (5). Leake (33) observed that an ericoid mycorrhizal fungus, *Hymenoscyphus ericae*, was able to metabolize some phytotoxic compounds *in vitro*. This ability of the fungus provides the host plant with added protection against toxins and allows the plant to grow in areas otherwise hazardous to the plant (33).

Herbicide Degradation. Several mycorrhizal fungi were screened for their ability to degrade 2,4-D and atrazine. Donnelly et al. (34) found that many of the fungi were capable of degrading these two chlorinated aromatic herbicides. The fungi were grown on medium containing ^{14}C-ring labelled 2,4-D or atrazine as the sole carbon source. After eight weeks of incubation, amounts of [14]CO_2 and ^{14}C incorporated into the fungal biomass was determined. The fungi were grown at two herbicide concentrations (1.0 and 4.0 mM) and three nitrogen concentrations (0, 1.0, and 10.0 mM). None of the cultures tested grew at 4.0 mM 2,4-D, but all of them grew at 1.0 mM 2,4-D. Both concentrations of atrazine also supported the growth of all the fungal cultures. In general, as the nitrogen concentration increased, the total herbicide degradation increased. In most cases, degradation consisted of incorporation of herbicide carbon into tissue and not mineralization. The ectomycorrhizal fungus *Rhizopogon vinicolor* and the ericoid mycorrhizal fungus *Hymenoscyphus ericae* were able to degrade 2,4-D. The two ericoid mycorrhizal fungi, *H. ericae* and *Oidiodendron griseum*, were able to degrade significant amounts of atrazine. The ectomycorrhizal fungi *Rhizopogon vulgaris* and *Gautieria crispa* were also able to metabolize atrazine. The ability to degrade these two herbicides was dependent on the fungus and the herbicide (34).

In a separate study, where sterile soil was colonized with mycorrhizal fungi, ^{14}C-ring labelled 2,4-D or atrazine was added at a concentration of 1.0 mM (35). The [14]CO_2 was trapped and monitored over a 30-day period. All of the mycorrhizal fungi tested mineralized more atrazine than 2,4-D. The ectomycorrhizal fungus *Gautieria crispa* mineralized the most atrazine, with the peak occurring between days 3 and 5, and increasing steadily after that. The ericoid mycorrhizal fungus *Oidiodendron griseum* mineralized the most 2,4-D, with the peak occurring between days 5 and 10 (35). Results of these studies showed that several of the mycorrhizal fungi which are able to degrade lignin were also able to degrade 2,4-D and atrazine (25,34,35).

Three commonly used herbicides were tested for their effect on selected ectomycorrhizal fungi (36-38). The herbicides tested were atrazine, hexazinone, and picloram. Axenic cultures of eleven ectomycorrhizal fungi were grown in liquid medium with the herbicide for 30 days. The medium was sampled periodically during this time period, and analyzed for disappearance of the parent compound and appearance of intermediate compounds (metabolites). The fungi were tested in media at two pH levels and with three C/N ratios. Robideaux et al. (37) showed that ten of the eleven ectomycorrhizal fungi were capable of removing hexazinone from the medium. The culture conditions and fungal species determined the amount of disappearance of the parent compound. The C/N ratio was a major factor involved in the amount of hexazinone disappearance (37). In a related study, Fowles et al. (38)

determined that the C/N ratio and pH affected the amount of hexazinone found immobilized in the fungal tissue. However, this effect was dependent on the fungal species. Of the ectomycorrhizal fungi tested, *Cenococcum geophilum* was the most effective at immobilizing the hexazinone (*38*). The analysis of the other two herbicides is not yet completed. These studies are consistent with the results reported by Donnelly et al. (*34,35*) that the ability to degrade herbicides appears to be dependent on the herbicide and the mycorrhizal fungus.

Several ectomycorrhizal fungi were also tested for their ability to degrade the herbicide chlorpropham (*39*). After one week of incubation with the substrate, the intermediate compound 3-chloroaniline was present in the medium of almost half of the cultures. In the samples with 3-chloroaniline, there was only a small amount of the parent compound found on the mycelium. The ectomycorrhizal fungi which degraded chlorpropham the best were *Hebeloma cylindrosporum*, *Suillus bellini*, and *Suillus variegatus*. This study showed that ectomycorrhizal fungi are capable of metabolizing herbicides, and the response of the fungus to the compound will vary among species (*39*).

PCB Degradation. Most recently, 21 mycorrhizal fungi were screened for their ability to metabolize PCBs (*40*). The fungi were incubated for five days in a liquid medium containing a mixture of known PCB congeners as the sole carbon source. The PCB congeners ranged from dichlorobiphenyl to hexachlorobiphenyl. After incubation, the PCB was extracted from the medium with hexane and analyzed using gas chromatography. Killed cells were used as controls and the % recovery of each PCB congener was compared between the living and dead cells. The fungi varied in their ability to metabolize the PCBs. Of the 19 total congeners present, the number of congeners degraded by the fungi ranged from zero to six. The lower chlorinated congeners were more readily degraded than the higher chlorinated congeners. Fourteen of the 19 fungi tested were able to metabolize some of the PCBs by at least 20%. The ectomycorrhizal fungi *Radiigera atrogleba* and *Hysterangium gardneri* were able to degrade over 80% of 2,2'-dichlorobiphenyl. Two ericoid mycorrhizal fungi, *Hymenoscyphus ericae* and *Oidiodendron griseum*, were not as effective at metabolizing PCB as the ectomycorrhizal fungi (*40*). The preliminary screening of mycorrhizal fungi showed that several were capable of metabolizing PCBs.

Summary

The limited research conducted to date with mycorrhizal fungi shows that this group of fungi have the ability to degrade a range of xenobiotics, similar in many respects to what has been reported for the white-rot fungus, *Phanerochaete chrysosporium* (*20,41*). The ability to degrade aromatic substrates depends upon the fungus and the substrate (*25,34,35,37,38,40*). Physiological conditions also have a significant effect on the rate of degradation (*25,34,37,38*). It should be emphasized that all of the work to date have been *in vitro* studies without the host plant. It is recognized that the mycorrhizal fungus may not metabolize these compounds to the same extent when growing with the host plant. *In vitro* studies with the intact symbiosis are now in progress, and actual field studies at contaminated sites are needed to determine the practical usefulness of this system. Both enzymatic and physiological studies completed with mycorrhizal fungi indicate that these fungi have the potential to play an important role in bioremediation of hazardous compounds in soil. The natural, sustained occurrence of abundant amounts of these fungi in association with plants make this plant-microbe system an ecologically sound approach to long-term bioremediation and ecological restoration of contaminated sites.

Literature Cited

1.   Garrett, S. D. *Soil Fungi and Soil Fertility*; 2nd edn.; Pergamon Press: Oxford, 1981; pp 150.
2.   Harley, J. L. *The Biology of Mycorrhiza*; 2nd edn.; Leonard Hill Books: London, 1969; pp 334.
3.   Harley, J. L.; Smith, S. E. *Mycorrhizal Symbiosis*; Academic Press, Inc.: London, 1983; pp 483.
4.   Harley, J. L. *Mycol. Res.* **1989,** *92,* 129-139.
5.   Paul, E. A.; Clark, F. E. *Soil Microbiology and Biochemistry;* Academic Press, Inc.: San Diego, CA, 1989; pp 273.
6.   Bae, K.-S.; Barton, L. L. *Appl. Environ. Microbiol.* **1989,** *55,* 2511-2516.
7.   Caldwell, B. A.; Castellano, M. A.; Griffiths, R. P. Abst. *8th North American Conference on Mycorrhizae*; 1990; p. 43.
8.   Caldwell, B. A., Griffiths, R. P.; Cromack Jr., K.; Castellano, M. A. Abst. *Soil Ecology Society Meeting*; 1991; p. 23.
9.   Englander, L. In *Methods and Principles of Mycorrhizal Research*; Schenck, N. C., Ed.; The American Phytopathological Society: St. Paul, MN, 1982, pp 11-13.
10.   Miller, Jr., O. K. In *Methods and Principles of Mycorrhizal Research*; Schenck, N. C., Ed.; The American Phytopathological Society: St. Paul, MN, 1982, pp 91-101.
11.   Trappe, J. M. *Bot. Rev.* **1962,** *28,* 538-606.
12.   Trappe, J. M.; Schenck, N. C. In *Methods and Principles of Mycorrhizal Research*; Schenck, N. C., Ed.; The American Phytopathological Society: St. Paul, MN, 1982, pp 1-9.
13.   Molina, R. J.; Trappe, J. M.; Strickler, G. S. *Can. J. Bot.* **1978,** *56,* 1691-1695.
14.   Rendig, V. V.; Taylor, H. M. . *Principles of Soil-Plant Interrelationships*; McGraw-Hill Publishing Company: New York, NY, 1989; pp 275.
15.   Hacskaylo, E. In *Ectomycorrhizae: Their Ecology and Physiology*; Marks, G. C.; Kozlowski, T. T. Eds.; Academic Press, Inc.: New York, NY, 1973; pp 207-230.
16.   Zhu, H.; Guo, D.-C.; Dancik, B. P. *Appl. Environ. Microbiol.* **1990,** *56,* 837-843.
17.   Leake, J. R. Abst. 8th *North American Conference on Mycorrhizae*; 1990; p. 177.
18.   Hutchison, L. J. *Can. J. Bot.* **1990,** *68,* 1522-1530.
19.   Leake, J. R.; Read, D. J. *New Phytol.* **1989,** *112,* 69-76.
20.   Aust, S. D. *Microb. Ecol.* **1990,** *20,* 197-209.
21.   Eaton, D. C. *Enzyme Microb. Technol.* **1985,** *7,* 194-196.
22.   Haider, K. M.; Martin, J. P. *Soil Biol. Biochem.* **1988,** *20,* 425-429.
23.   Hammel, K. E. *Enzyme Microb. Technol.* **1989,** *11,* 776-777.
24.   Schreiner, R. P.; Stevens Jr., S. E.; Tien, M. *Appl. Environ. Microbiol.* **1988,** *54,* 1858-1860.
25.   Donnelly, P. K.; Entry, J. A.; Cromack Jr., K.; Crawford, D. L.; Caldwell, B. A. *Microb. Ecol.* **1993,** Submitted.
26.   Entry, J. A.; Donnelly, P. K.; Cromack, Jr., K. *Biol. Fertil. Soils* **1991,** *11,* 75-78.
27.   Haselwandter, K.; Bobleter, O.; Read, D. J. *Arch. Microbiol.* **1990,** *153,* 352-354.
28.   Trojanowski, J.; Haider, K.; Huttermann, A. *Arch. Microbiol.* **1984,** *139,* 202-206.

29. Cromack Jr., K.; Sollins, P.; Graustein, W. C.; Speidel, K.; Todd, A. W.; Spycher, G.; Li, C. Y.; Todd, R. L. *Soil. Biol. Biochem.* **1979,** *11,* 463-468.
30. Fogel, R.; Hunt, G. *Can. J. For. Res.* **1979,** *9,* 245-256.
31. Cromack Jr., K.; Fichter, B. L.; Moldenke, A. M.; Entry, J. A.; Ingham, E. R. *Agric. Ecosyst. Environ.* **1988,** *24,* 161-168.
32. Entry, J. A.; Rose, C. L.; Cromack Jr., K. *Soil. Biol. Biochem.* **1991,** *23,* 285-290.
33. Leake, J. R. In *Mycorrhizae in the Next Decade: Practical Applications and Research Priorities*; Sylvia, D. M.; Hung, L. L.; Graham, J. H. Eds.; Institute of Food and Agricultural Sciences, University of Florida: Gainesville, FL, 1987, p. 332.
34. Donnelly, P. K.; Entry, J. A.; Crawford, D. L. *Appl. Environ. Microbiol.* **1993,** *59,* 2642-2647.
35. Donnelly, P. K.; Entry, J. A.; Rygiewicz, P. T.; Crawford, D. L. *Appl. Environ. Microbiol.* **1993,** Submitted.
36. Robideaux, M. L.; Wickster, P. V.; DeWhitt, K.; Spriggs, J. W.; Rygiewicz, P. T. Abst. *North American Conference on Mycorrhizae;* 1990, p. 250.
37. Robideaux, M. L.; Fowles, N. L.; King, R. J.; Spriggs, J. W.; Rygiewicz, P. T. Abst. *Society of Environmental Toxicology and Chemistry*; 1991, p. 175.
38. Fowles, N. L.; Robideaux, M. L.; King, R. J.; Spriggs, J. W.; Rygiewicz, P. T. Abst. *Society of Environmental Toxicology and Chemistry*; 1991, p. 176.
39. Rouillon, R.; Poulon, C.; Bastide, J.; Coste, C. M. *Agric., Ecosystems Environ.,* **1989,** *28,* 42-424.
40. Donnelly, P. K.; Fletcher, J. S. *Environ. Sci. Tech.* **1993,** Submitted.
41. Field, J. A.; de Jong, E.; Feijoo-Costa, G.; de Bont, J. A. M. *Tibtech.* **1993,** *11,* 44-49.

RECEIVED February 15, 1994

INDUSTRIAL CHEMICALS

Chapter 9

Dehalogenation of Chlorinated Phenols During Binding to Humus

Jerzy Dec and Jean-Marc Bollag

Laboratory of Soil Biochemistry, Center for Bioremediation
and Detoxification, Pennsylvania State University,
University Park, PA 16802

Oxidoreductive enzymes, occurring in the soil environment, can
catalyze oxidative coupling reactions of chlorinated phenols. The
free radicals generated during these reactions can couple to each
other or to humic acid. If the unpaired electron of a free radical is
located at an aromatic carbon which is substituted by a chlorine
atom, dehalogenation can occur during the coupling reaction. The
patterns of dehalogenation during the binding to humus are
comparable to those observed when free radicals couple to each
other, which confirms that both reactions are controlled by the
same mechanism. The occurrence of dehalogenation during
binding provides direct evidence for the formation of covalent
bonds between chlorinated phenols and humic acid. Covalent
binding is considered the strongest type of bonding   between
xenobiotics and soil organic matter, and therefore - together with
the dehalogenation process - is a desired reaction for the removal
of chlorinated phenols from the environment.

There is growing evidence that, due to increased microbial activity, the
degradation rate of hazardous organic compounds in the rhizosphere is greater
than in root-free soil (*1,2*). In addition to degradation, xenobiotics deposited in
soil may also undergo synthetic processes, such as polymerization or binding to
humus (*3*). Numerous studies demonstrate that, like degradation, binding of
xenobiotics to soil depends on the activity of soil microorganisms; in sterile soils
essentially no significant binding is observed (*4*). Therefore, it can be expected
that in the terrestrial environments of increased microbial activity, such as
rhizosphere soil, the binding processes may also intensify.

Chlorinated phenols can be bound to soil organic matter through an oxidative
coupling reaction that may be mediated by various oxidoreductive enzymes such

0097–6156/94/0563–0102$08.00/0

as peroxidases, laccases, and tyrosinases (5). Oxidoreductase activity has been found in many bacteria and fungi occurring in soil. Peroxidase activity, for instance, was detected in the culture filtrate of the soil fungus *Geotrichum candidum* and in two *Bacillus* spp., a *Pseudomonas* sp. and an *Arthrobacter* sp. (6). Two species of *Aspergillus* and a species of *Streptomyces* isolated from soil exhibited similar peroxidase activity. Intracellular and extracellular laccases were found in many fungi (e.g. Agaricales, *Aspergillus fumigatus*, *Rhizoctonia praticola*, *Trametes versicolor* and others) and in actinomycetes like *A. albocrustosus*, *A. galbus*, and *Streptomyces* spp. (7-9). Tyrosinase activity was detected in fungi such as Agaricales, *Aspergillus nidulans*, brown-rot fungi, *Psalliota arvense*, and *Russula nigricans* (10,11), and in various actinomycetes (8). Both peroxidase and laccase activity were also detected in soil extracts, apparently as a result of the production of the extracellular oxidoreductases (12-15).

Rouatt et al. (16) reported that the number of bacteria, fungi, and actinomycetes in 1 g of rhizosphere soil may be 23, 12 and 7 times greater, respectively, than that in non-rhizosphere soil. Accordingly, an increase in the oxidoreductase activity can be expected in the root zone, leading to the enhanced participation of oxidative coupling reactions in the overall transformation of xenobiotics taking place in soil. The intensification of oxidative coupling in the rhizosphere may also be due to the presence of some oxidoreductases in the plant root exudates (17).

In recent studies (18), we demonstrated that binding of chlorinated phenols to humic acid is accompanied by the release of chloride ions from the substrates. Such a dehalogenation reaction could easily be mistaken for a symptom of biodegradation processes, especially if it is detected in the rhizosphere, a zone of increased microbial activity. In reality, as is demonstrated below, the role of soil microbiota is often limited to providing oxidoreductases, whereas the immediate cause of dehalogenation can be the oxidative coupling reaction through which chlorophenols bind to humus.

Investigatory Approach

The occurrence of chloride release during oxidative coupling processes is well documented (19-23). However, in previous research, release of chloride ions from chlorinated phenols was demonstrated only during polymerization of the substrates in the presence of oxidoreductive enzymes; dehalogenation during other oxidative processes was not investigated. As a result, it was accepted that the enzymes caused the dehalogenation effect (19-23). However, it was our hypothesis that the oxidative coupling reaction - and not enzyme activity - is the immediate cause of chloride release (18). In order to confirm this hypothesis, we initiated studies on dehalogenation occurring during both polymerization and binding of chlorinated phenols to humic acid in the presence of various oxidoreductases (horseradish peroxidase, *Trametes versicolor* laccase, tyrosinase).

Because oxidative coupling can be mediated not only by oxidoreductive enzymes but also by inorganic catalysts, we also investigated dehalogenation of chlorinated phenols during polymerization mediated by birnessite. Five chlorinated phenols were investigated: 2-chlorophenol (2-CP); 3-chlorophenol (3-CP); 4-chlorophenol (4-CP); 2,4-dichlorophenol (DCP); and 2,4,5-trichlorophenol (TCP). The reaction mixtures were analyzed by high-performance liquid chromatography for remaining substrates and coulometric titration for released chloride ions (18). Radiocounting was applied to measure binding of the ^{14}C-labeled substrates to humic acid. The percentages of substrate transformed (% Tr) and chloride released (% Cl) were used to determine the stoichiometry of dehalogenation. Dividing the % Tr by the % Cl and by the number of chlorine atoms on the benzene ring, the dehalogenation number (DN) was calculated; this represents the number of substrate molecules per one released chloride ion (18).

Results and Discussion

The results confirmed the hypothesis that oxidative coupling reaction is responsible for dehalogenation of chlorinated phenols (18). Release of chloride ions was observed during both polymerization of chlorophenols mediated by birnessite and binding of chlorophenols to humic acid in the presence of horseradish peroxidase (Table 1). The percentages of substrate transformed and chloride released for horseradish peroxidase-mediated polymerization differed from those determined for birnessite, but the respective dehalogenation numbers were very similar (Table 1).

It is known that in the presence of horseradish peroxidase or birnessite chlorinated phenols are first oxidized to free radicals (5,24). After the free radicals are generated, coupling is completed without further involvement of the catalyst. Various products may be formed during coupling due to the delocalization of the unpaired electron and the occurrence of the free radical in several resonance forms (25). Our experimental DN values of dehalogenation number indicated that the location of the unpaired electron determines whether or not dehalogenation may take place (18). In the case of chlorophenols, the lone electron may be located on the oxygen atom or on the aromatic ring in the *ortho* or *para* position. If the *ortho* or *para* position with the unpaired electron is substituted with the chlorine atom, the coupling of such a radical should result in a dehalogenation event.

According to this scheme, 10 chloride ions can be released from 20 molecules of DCP involved in a free radical dimerization presented in Figure 1. The calculated DN value for such a model (20/10) is 2, which is in agreement with DN values obtained experimentally (Table 1). A similar agreement was obtained for dehalogenation of other chlorophenols (Table 1), confirming the hypothesis that the release of chloride ions is related to the location of the unpaired electron in the free radical molecule (18).

Table 1.   The percent transformation (%Tr) and dehalogenation (%Cl) of chlorophenols during various oxidative coupling processes and the respective dehalogenation numbers (DN) (adapted from ref. 18)

Chlorophenols	Polymerization with horseradish peroxidase			Polymerization with birnessite			Binding to humic acid		
	%Tr	%Cl	DN	%Tr	%Cl	DN	%Tr	%Cl	DN
2-Chlorophenol	14.0	2.3	6.1	81.6	12.7	6.4	100.0	19.7	5.1
3-Chlorophenol	10.0	0.2	50.0	65.5	1.2	54.5	97.3	6.7	14.5
4-Chlorophenol	20.0	6.9	2.9	80.3	25.0	3.2	93.4	41.4	2.3
2,4-Dichlorophenol	94.0	24.0	2.0	92.0	22.0	2.1	100.0	37.9	1.3
2,4,5-Trichlorophenol	83.5	12.5	2.2	84.3	14.0	2.0	100.0	29.8	1.1

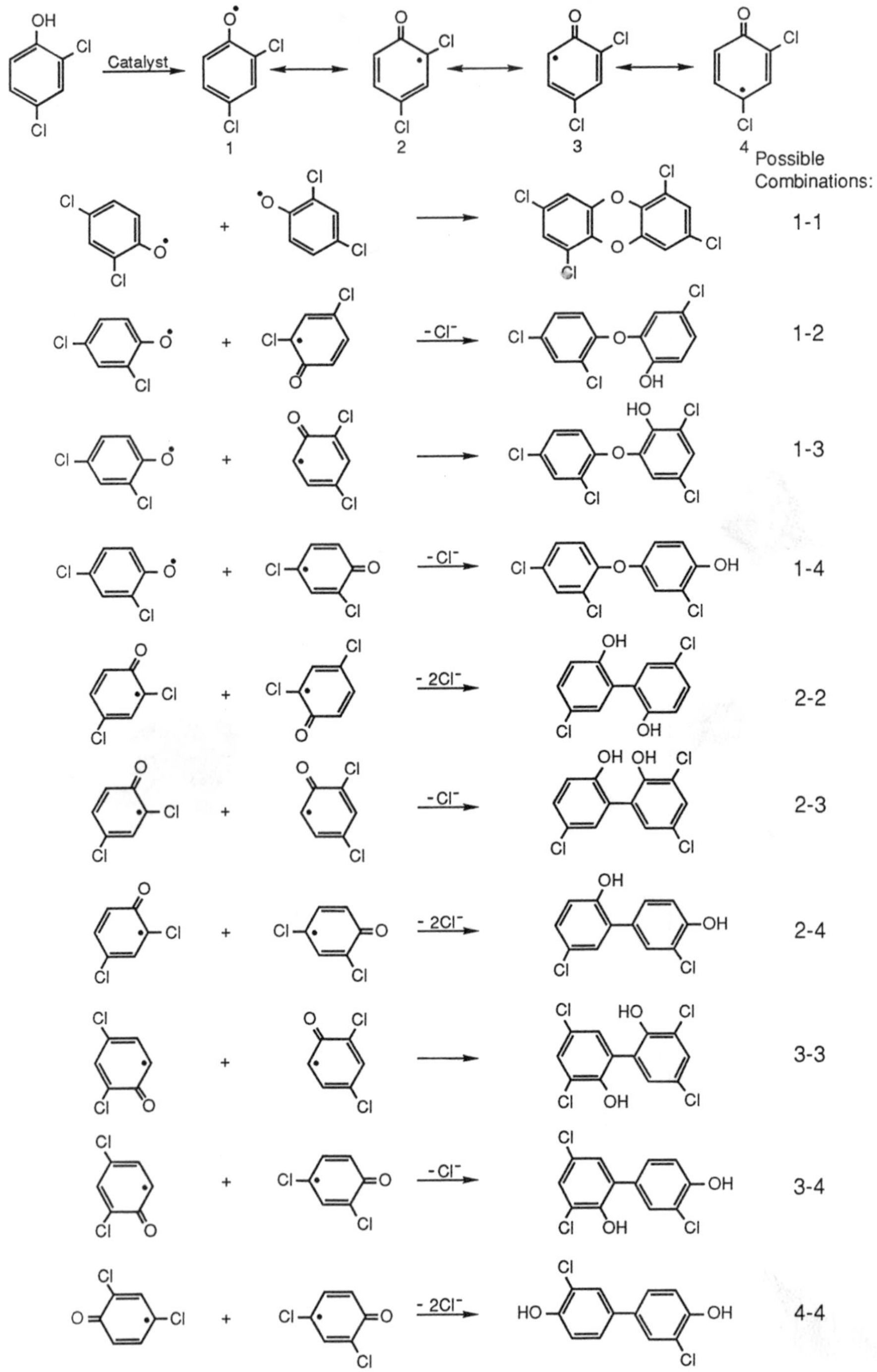

Figure 1. Proposed coupling reactions between free radicals generated during catalytic oxidation of 2,4-dichlorophenol (adapted from ref. 18).

Because the unpaired electron never occurs at the *meta* position, release of the chlorine atom from 3-CP is not expected. Accordingly, only small amounts of chloride ion were detected in the reaction mixtures containing this substrate (Table 1), and DN values amounted to 50.0 for horseradish peroxidase and 54.5 for birnessite. This limited dehalogenation can be ascribed to the release of chloride ions from oligomer radicals. In oligomers of 3-CP, the chlorine atoms that were previously located at the *meta* position, become *ortho* or *para* substituted, i.e. attached to the carbon that can host the unpaired electron acquired during radical transfer. Apparently, a fraction of these chlorines is subject to release.

Release of chloride ions during polymerization of chlorophenols was also observed in the presence of tyrosinase and the laccase from the fungus *Trametes versicolor*. The dehalogenation numbers for the laccase were similar to those determined for horseradish peroxidase and birnessite, but those resulting from incubations with tyrosinase were different. The differences reflect the fact that tyrosinase-mediated oxidation of chlorophenols leads to the formation of *ortho*-quinones (*26*), not the generation of free radicals, and oxidative coupling is governed by a different mechanism involving nucleophilic attack of the *ortho*-quinones by phenoxide anions. Release of chloride ions resulting from the free radical coupling was also observed during incubation of chloroanilines with horseradish peroxidase.

Dehalogenation patterns determined for horseradish peroxidase-mediated binding of chlorophenols to humic acid indicate that the process of binding is also governed by the free radical mechanism (*18*) (Table 1). No precipitation was observed during incubations with humic acid, indicating that no polymerization occurred, and the removal and dehalogenation of the substrates was entirely due to binding. This observation was confirmed by the fact that, for chlorophenols labeled with ^{14}C, most radioactivity (88.7% for 4-CP, 76.7% for DCP, and 81.9% for TCP) was found in the humic acid fraction after its precipitation with HCl (18).

Using electron spin resonance spectrometry (ESR), previous studies have determined that stable free radicals occur in humic acids (*27*). However, these free radicals apparently did not significantly contribute to the observed enzyme-mediated binding. According to the estimates of Schnitzer and Skinner (*27*), for fulvic acid, one free radical occurs for as many as 44,000 carbon atoms and 30,000 oxygen atoms. This density of stable free radicals seems far too low to account for the complete removal of chlorophenols that was observed during the binding reaction in our study (Table 1). Therefore, it is assumed that the required amounts of free radicals are created in humic acid during incubation with enzyme through radical transfer (*18*).

The experimental DN values for binding to humic acid are lower (i.e. enhanced dehalogenation) than those determined for polymerization (Table 1). This is partly due to the reduced number of possible couplings. During polymerization, free radicals of 2-CP, 3-CP, DCP, and TCP could couple to each other in ten different ways (Figure 1), and those of 4-CP in six different ways. In the case of binding, the free radicals can couple only to humic acid and the number of possible couplings for 2-CP, 3-CP, DCP and TCP is reduced to four, and to three for 4-CP. In other words, the number of possible couplings is equal to the number of free radicals possessed by a given substrate (*18*). The enhanced dehalogenation during binding may also be related to coupling of chlorophenols to already bound substrates. The experimental DN value for binding of 3-CP (14.5) that indicates the possibility of such a coupling is about three times less than that obtained for horseradish-mediated polymerization (50.0). Similarly, reduced DN values were obtained for binding of 4-CP, DCP, and TCP (2.3, 1.3, and 1.1, respectively), by comparison to the polymerization DN values (2.9, 2.0, and 2.2, respectively).

The coupling of chlorophenols to already bound substrates may explain our earlier findings from ^{13}C-NMR spectroscopy that indicated formation of a dimeric moiety (Substructure I) during enzymatic binding of ^{13}C-labeled DCP to humic acid (*28*).

Substructure I

This dimeric moiety is formed by the coupling between carbons at the *ortho* (C-6) and the *meta* position (C-3'). Such coupling is impossible unless the *meta* position changes its status to *para* position after coupling of the DCP free radical to humic acid (HA).

The *meta* position carbon that assumed the *para* position can host the unpaired electron acquired through the radical transfer and can be coupled with another DCP free radical to form a complex of humic acid with the dimeric moiety identified in the ^{13}C-NMR studies.

Various interactions may be responsible for adsorption or binding of xenobiotics to humic substances, such as covalent bonds, van der Waal's forces, H-bonding, ion exchange, hydrophobic bonding, ligand exchange, or entrapment in the molecular net of humus (*3,4*).   However, except for covalent bond formation, none of these interactions results in dehalogenation of chlorophenols. Therefore, the release of chloride ions constitutes direct evidence for the formation of covalent bonds between chlorophenols and humic acid during enzyme-mediated binding.   Due to the resonance of the unpaired electron, both carbon-oxygen and carbon-carbon linkages may be formed.   These findings are in agreement with previous investigations on binding by means of mass spectroscopy, radiocounting of ^{14}C-labeled substrates, and ^{13}C-NMR spectrometry (*5,28*).

Conclusions

Dehalogenation of organic chemicals in terrestrial and aqueous environments is a common phenomenon.   Usually, it is caused by the activity of microorganisms (*29*), but can also be due to physicochemical factors, e.g. photolysis (*30*).   Microbial dehalogenation takes place under both aerobic (*31*) and anaerobic conditions (*32*).   In many cases, dehalogenation constitutes an initial step leading to further decomposition of organic compounds (*31*).   On the other hand, the loss of chlorine may be a result of the ongoing breakdown initiated by other factors (*33*).   Under certain conditions, dehalogenation may have no immediate consequence, except the decreased toxicity of the resulting dehalogenated compound (*34*).

Release of chloride ions during oxidative coupling processes constitutes another possible type of dehalogenation (*18*).   Like the other types, it may be more intensive in the rhizosphere.   Covalent binding is considered the strongest type of bonding between xenobiotics and soil organic matter, and therefore - together with the dehalogenation process - is a desired reaction for the removal of chlorinated phenols from the environment (*5*).   The contribution of oxidative coupling to the overall dechlorination taking place in soil can be estimated by measuring the extent of the synthetic processes.   This contribution appears to be significant, taking into account that, typically, as much as 20-70% of the chemical entering the terrestrial system may become bound to soil (35). The DN value of dehalogenation number, applied in our study to elucidate the mechanism of oxidative coupling, may prove helpful in differentiating between various types of dehalogenation occurring in terrestrial systems, both in the rhizosphere and root-free soil. At the same time, it may prove useful for evaluating bioremediation procedures based on oxidative coupling and other mechanisms.

Literature Cited

1. Klein, D.A. In *Encyclopedia of Microbiology;* Academic Press, Inc., **1992**; Vol. 3, pp. 565-572.
2. Walton, B.T.; Anderson, T.A. *Curr. Opin. Biotechnol.* **1992**, *3, 267-270.*
3. Bollag, J.-M.; Loll, M.J. *Experientia* **1983**, *39*, 1221-1231.
4. Khan, S.U. *Res. Rev.* **1982**, *84*, 1-25.
5. Bollag, J.-M. *Environ. Sci. Technol.* **1992**, *26*, 1876-1881.
6. Bordeleau, L.M.; Bartha, R. *Can. J. Microbiol.* **1972**, *18*, 1865-1971.
7. Matsubara, T.; Iwasaki, H. *Bot. Mag. (Tokyo)* **1972**, *85*, 71-83.
8. Krasil'nikov, N.A.; Zenova, G.M.; Bushueva, O.A. *Mikrobiologiya* **1974**, *43*, 606-608.
9. Bollag, J.-M.; Sjoblad, R.D.; Liu, S.-Y. *Can. J. Microbiol.* **1979**, *25*, 229-233.
10. Bourquelot, E.; Bertrand, G. *Soc. Biol.* **1895**, *47*, 582-584.
11. Safe, S.; Ellis, B.E.; Hutzinger, O. *Can. J. Microbiol.* **1976**, *22*, 104-106.
12. Galstyan, A.S. *Dokl. Akad. Nauk. Arm. SSR.* **1958**, *26*, 285-288.
13. Kozlov, K.A. *Folia Microbiol.* **1964**, *9*, 145-149.
14. Bartha, R.; Bordeleau, L. *Soil Biol. Biochem.* **1969**, *1*, 139-143.
15. Bollag, J.-M.; Chen, C.-M.; Sarkar, J.M.; Loll, J.M. *Soil Biol. Biochem.* **1987**, *19*, 61-67.
16. Rouatt, J.W.; Katznelson, H.; Payne, T.M. *Soil Sci. Am. Proc.* **1960**, *24*, 271-273.
17. Spalding, B.P.; Duxbury, J.M.; Stone, E.L. *Soil Sci. Soc. Am. Proc.* **1975**, *39*, 65-70.
18. Dec, J.; Bollag, J.-M. *Environ. Sci. Technol.* (submitted).
19. Lyr, V.H. *Phytopathol. Z.* **1963**, *47*, 73-83.
20. Hammel, K.E.; Tardone, P.J. *Biochemistry* **1988**, *27*, 6563-6568.
21. Minard, R.D.; Liu, S.-Y.; Bollag, J.-M. *J. Agric. Food Chem.* **1981**, *29*, 250-253.
22. Dec, J.; Bollag, J.-M. *Arch. Environ. Contam. Toxicol.* **1990**, *19*, 543-550.
23. Roy-Arcand, L.; Archibald, F.S. *Enzyme Microb. Technol.* **1991**, *13*, 194-203.
24. Shindo, H.; Huang, P.M. *Soil. Sci. Soc. Am. J.* **1984**, *48*, 927-934.

25. Nonhebel, D.C.; Walton, J.C. *Free-radical chemistry; structure and mechanism;* Cambridge University Press: London, 1974.
26. Malmström, B.G.; Rydén, L. In *Biological Oxidations;* Singer, T.P., Ed.; Interscience Publishers: New York, 1968, pp. 415-438.
27. Schnitzer, M.; Skinner, S.I.M. *Soil Sci.* **1969**, *108*, 383-390.
28. Hatcher, P.G.; Bortiatynski, J.M.; Minard, R.D.; Dec, J.; Bollag, J.-M. *Environ. Sci. Technol.* **1993**, *27*, 2098-2103.
29. Reinecke, W.; Knackmuss, H.-J. *Appl. Environ. Microbiol.* **1984**, *47*, 395-402.
30. Boule, P.; Guyon, C.; Lemaire, J. *Chemosphere* **1984**, *13* , 603-612.
31. Kiyohara, H.; Takizawa, N.; Uchiyama, T; Ikarugi, H.; Nagao, K. *J. Ferment. Bioeng.* **1989**, *67*, 339-344.
32. Hendriksen, H.; Ahring, B.K. *Biodegradation* **1993**, *3*, 399-408.
33. Steiert, J.G; Crawford, R.L. *Trends Biotechnol.* **1985**, *3*, 300-305.
34. Adriaens, P.; Kohler, H.-P.E.; Kohler-Staub, D.; Focht, D.D. *Appl. Environ. Microbiol.* **1989**, *55*, 887-892.
35. Calderbank, A. *Rev. Environ. Contam. Toxicol.* **1989**, *108*, 71-103.

RECEIVED February 22, 1994

Chapter 10

Alfalfa Plants and Associated Microorganisms Promote Biodegradation Rather Than Volatilization of Organic Substances from Ground Water

Lawrence C. Davis[1], N. Muralidharan[2], V. P. Visser[3],
C. Chaffin[3], W. G. Fateley[3], L. E. Erickson[2], and R. M. Hammaker[2]

Departments of [1]Biochemistry, [2]Chemical Engineering, and [3]Chemistry, Kansas State University, Manhattan, KS 66506

Bacteria have the capacity for bioremediation of both volatile and non-volatile organic compounds. Plants support a rhizosphere microflora, enhance soil microbial populations and may also be able themselves to metabolize some hazardous organic compounds. Plants move large amounts of water by transpiration and for volatile compounds, intersystem transfer by transpiration might occur when plants are exposed to such materials. These possibilities were tested experimentally. A tank with a channel width of 10 cm and a depth of 35 cm was used for plant growth. Alfalfa plants were supplied with a subsoil water source saturated (c. 500ppm) with toluene or containing c. 500 ppm phenol. Sampling wells were used to monitor the depth of the water table and concentrations of toluene and phenol in the ground water.

The toluene concentration in the ground water remained constant or decreased slightly during passage of water through the tank although there was an average net input of >300 mg toluene per day to the system. To measure toluene losses through volatilization or transpiration, the tank was covered with an aluminum and glass enclosure and monitored using Fourier transform infra-red (FTIR) spectroscopy. The gas phase toluene was below the limit of detection by FTIR (c. 250 ppb v/v) while the expected accumulation of toluene based on water evapotranspiration rate was >50 ppm v/v increase in the gas phase concentration per hour. Thus, only a small to nil fraction of the input toluene arrived in the gas phase.

With phenol, the concentration in the aqueous phase decreased during passage through the tank. Phenol is much less volatile than water and it was undetectable in the gas phase. The amount of input contaminants was about 100 fold above the amount that could be physically adsorbed to soil organic matter in this system. The observed mass balance suggests that effective degradation of toluene and phenol occurs in this system and that the potential intersystem transfer of volatile organics by plant transpiration is not a problem, at least with adapted plants.

Bioremediation efforts have generally depended on microorganisms alone but plants may make significant additional contributions to the process.   Plants are just beginning to be used specifically in this role (*1-3*), although they have been used in mine waste reclamation for decades.   We are particularly interested in remediation of ground water contaminated by organic compounds.   A recent review (*4*) describes the current status of and potential applications of plant assisted bioremediation.   Plants provide a cost-effective way to "pump" contaminated water out of the ground, at least in less humid regions where potential evapotranspiration exceeds precipitation.   They also provide input of readily available supplemental organic carbon to enhance the growth of root-associated microorganisms.   This, in turn, increases the total population of microorganisms in the soil.

We previously modeled the plant-assisted bioremediation strategy for benzene and atrazine (*5,6*).   Plants reduce the off-site transfer of the hazardous substance, primarily by reducing downward percolation of water, but also by withdrawing ground water.   The relative withdrawal of ground water depends on the water demand of the plant species chosen, precipitation, and potential evapotranspiration.   Degradation of organic contaminants depends on microbial biomass, which, in turn, is strongly dependent on carbon input from plant root exudate.   In the model (*5,6*), as in real systems (*4*), total microbial biomass in the soil is strongly correlated with root density in the soil.   To validate the model, we have constructed a system that allows direct measurement of toluene or phenol moving into and out of the saturated soil phase, into the vadose zone and potentially into the atmosphere.

Earlier work by McFarlane et al. (*7*) had shown that soil could serve as a sink for benzene from the vapor phase.   In their studies, it was concluded that microorganisms in the soil, associated with the plant root systems, were responsible for the disappearance of benzene.   Sterilized soil was not a sink for benzene and unplanted soil was much less effective than soil which had plant roots present. When a highly active rhizosphere microbial population is present, degradation of the hazardous organic substrate may be relatively rapid.   Walton and Anderson (*2*) found that degradation of trichloroethylene was greater in soil from the rhizosphere of plants growing in a contaminated site than with contaminated soil in which plants were not growing.   Microorganisms, either closely or loosely associated with plant roots, are expected to enhance degradation of toluene and phenol, as there are many bacteria capable of degrading these compounds.   Microbially enhanced bioremediation of these compounds is considered a standard technique (*8*).

The use of plants is an extension and enhancement of the intrinsic bioremediation capabilities of the microbial community as discussed in a recent National Research Council book (*8*), rather than a fully engineered approach.   A crucial question in plant assisted bioremediation is whether there is significant transfer of organic contaminants from soil to the plant and atmosphere.   Briggs et al. (*9*) showed in short term tests (2-4 days) that barley plants will take up dissolved organics into their transpiration stream.   Compounds in homologous series (phenylureas and methylcarbamoyloximes) with varied octanol/water partition coefficients were tested. There was considerable variation between compounds, but those with a partition coefficient of about 100 ($\log K_{ow}$ = 2) had a maximum transpiration stream concentration factor (TSCF) of about 0.8.   That is, such compounds were about 80% as concentrated in the transpiration stream as outside the root of the plant. Greater or lesser hydrophobicity yielded less effective uptake. The phenylureas consistently had lower TSCF values than the O-methylcarbamoyloximes.

McFarlane et al. (*10*) using an elegantly designed hydroponic system, showed that compounds in different chemical classes but having similar octanol/water partition coefficients may behave in quite different fashion depending on the species of plant tested.   Phenol appears to be almost entirely immobilized in the roots of soybeans, while bromacil is rapidly translocated to and accumulates in the leaves. Similarly, Briggs et al. (9) cited a number of examples from the literature where the TSCF deviated greatly, to the low side, from their relatively simple predictive curve. We may not be able to predict *a priori* whether a particular compound is translocated or immobilized by a particular species of plant. In addition, use of a non-sterile planted soil as the support medium may give different results from those observed in hydroponic systems because there are many more potential compartments for contaminant to partition into and there may be degradative losses in the soil (11).

An active plant consumes relatively large amounts of water as an inescapable consequence of photosynthesis (4-6). Adapted microorganisms associated with the roots may carry out degradation of dissolved contaminants. Thus the plant may compete with its associated microorganisms by taking up contaminants from the ground water. Or, if the microbial community is large enough and effectively adapted to the contaminant in question, the plant may serve primarily to enhance active transport of the contaminant to the microorganisms. The relative importance of these competing processes is addressed in the present research.

Boersma et al. (11) showed that accumulation of bromacil in plants increased in proportion to transpiration rate. In the absence of precipitation, an active plant significantly increases the net water flux from water table through the vadose zone to air. Removal of water increases the volume of the gas phase, which allows increased oxygen diffusion through the soil, potentially enhancing aerobic processes. In saturated soils, the root systems of some species may also supply some oxygen for microbial consumption (4). Plants thus may expose microorganisms to increased fluxes of both gas and water, which contain the volatilized or solubilized contaminant. Adapted microbial populations will probably consume most of the contaminant before it is taken up by the roots, because typical K_m values for microbial metabolism of compounds such as phenol are fairly low, in the range of 100 µg/L (12), several thousand times below the concentrations being used in the experiments described here.

The work of Briggs et al. (9), and Boersma et al. (11) followed uptake of contaminants for relatively short periods of time. Soil microorganisms able to degrade a compound increase following exposure (13). Thus the potential uptake of organic compounds by a plant and transfer to the gas phase which is predicted from their work, might be decreased in an adapted system, where associated microorganisms could degrade the contaminant prior to uptake by the plant. This is an important consideration because intermedia transfer of contaminants is a concern in any remediation process. The present research directly addresses that question.

Materials and Methods

Apparatus for treatment of plants. A tank divided into two identical halves with inlet and outlet ports was used for plant growth. The U-shaped folded channel in each half of the tank was 1.8 m long, 10 cm wide and 35 cm deep. It was filled with Kansas river sand/silt from the Sweet tract, near the Riley County landfill where transport of organics in ground water is a concern. Details of construction and soil packing were previously described (14). Four sampling wells were placed in each half chamber at the time of soil packing. Each well consisted of a coarse glass frit on a gas dispersion tube with the tube lengthened to 38 cm. Concentrations of toluene and phenol in the ground water were monitored through these sampling wells which also allowed determination of the depth of the water table. With reasonable flow rates and a water table depth of 25 cm from the surface, the 1.8 m path would provide a mean residence time of 1-3 weeks for input liquid. Evaporation rate greatly influences the mean residence time of the liquid effluent. The limiting case is when inlet flow just equals evapotranspiration and residence time becomes indefinitely long. The surface area of soil for each half chamber was about 0.18 m^2, while the area available for plant top growth was slightly larger (c. 0.2 m^2) because plants could grow above the dividing walls. The entire chamber was placed within a fume hood to insure that any possible escape of volatiles did not violate regulations for air quality within the laboratory.

To monitor possible toluene or phenol output through volatilization or transpiration, the tank was covered with a gas-tight enclosure 26 cm high. On three sides and part of the fourth, steel covered with aluminum flashing material was used to provide a light reflecting surface that would not adsorb organic compounds. The top of the enclosure was covered by ordinary window glass and sealed with clear tape around the edges. Illumination for the plants was provided by six 40 watt fluorescent lights 40 cm above ground level. During monitoring periods, the front open space, about 43 cm wide, was closed with a piece of window glass and sealed with tape. Internal air circulation was assured by using a small fan inside the enclosure.

Soil analysis and plant growth. The soil used in these studies consists of medium and fine sand with a little silt and low organic matter (0.3 to 1.3 % depending on depth)(*14*). Effectively nodulated alfalfa seedlings were planted in pairs at 10-cm intervals along the length of the flow path and the whole tank was top watered once with 10 L Wacek's medium minus N (*15*) to give the plants essential nutrients. A subsurface flow of water was established using water saturated with toluene or having c. 500 ppm (w/v) of phenol (500 μL/L of 93%, liquified, phenol). The water table was maintained at 20-30 cm and the gradient in the central 1 m of the flow path was about 1 cm/m.

Analysis of aqueous phase organics. Fresh representative samples of the ground water were taken from the sampling wells and the concentration of phenol or toluene at the entry and exit ports was also measured. After drawing out all the liquid in a well and allowing it to refill for 5-10 minutes, a measured quantity of liquid was withdrawn into a long nylon tube attached to a 1-mL syringe. The 1-mL liquid sample was added to organic extractant in a 50-mL tube closed with a ground glass stopper. For toluene extraction, 2 mL of heptane was used. For phenol, 2 mL n-octanol was used with 5 mL of 0.1 M phosphate buffer added to assure that the phenol remained unionized. A vortex mixer was used to facilitate partitioning of the toluene or phenol into the organic phase. After separation of phases, a portion of the organic phase was transferred to a quartz cuvette. The absorbance, at 262 nm for toluene or 265 nm for phenol, was determined in a DU-2 spectrophotometer. Calibration standards were prepared, and extracted as for the unknowns. For some samples, spectra of the entire UV region from 200 - 300 nm were determined using a Hitachi Model U-3210 recording spectrophotometer.

Analysis of gas phase organics. Details of instrumentation are given in Davis et al. (*14*). Gas phase concentrations of toluene were monitored after closing the growth chamber to give an aboveground gas volume of 147 L. The walls of the upper enclosure had two mirrors mounted near the top to allow a beam of light to enter from the front of the chamber, pass to the back where it was reflected to the opposite end, return to the first mirror and exit at the front. Total path within the chamber was 2.44 m. Entry and exit apertures, covered by KBr plates, were slightly offset to align with the source and entry of a MIDAC FTIR. Chilling coils with refrigerant at a nominal -10°C circulated near the base of the upper enclosure to increase transpiration and decrease humidity near the KBr windows. Methane (10 mL) was used as a freely diffusible inert gas for measuring leakage out of the upper enclosure. In earlier experiments, the half-time for equilibration was about 1 hour, while in later experiments it was three hours or greater.

FTIR Instrument calibration. Several methods were used to test the sensitivity of the MIDAC detection system. In one case, toluene was applied to a piece of paper within the chamber via a long syringe needle. It was allowed to evaporate and the spectrum measured. In another case, water saturated with toluene was siphoned from the inlet reservoir into a container and then directly applied rapidly to the surface of the chamber soil. The chamber was then immediately sealed, and the increase in toluene level in the gas phase was monitored. In a third case, water similarly saturated with toluene was applied to dry soil from the same site placed in four 23 x 33-cm pyrex baking trays which had been arranged in the chamber after the plants had been cut back to 5 cm. Finally, the potential evaporation of toluene was enhanced by flooding the chamber from below to within 10 cm of the surface with water from the inlet reservoir, simply by elevating the pressure head for a few hours. For phenol, only the direct addition of liquid phenol on filter paper was used in sensitivity tests.

Respiration rate estimation. Two different methods were used to measure respiratory CO_2. The simpler method was to place closed cylindrical containers between the plants, with their lower edge pressed into the soil 1-2 cm. Seven containers of 7-cm diameter and 850-mL volume were placed at intervals between the plants. Gas samples of 1 mL were taken by syringe, through septa placed in the tops of the containers. Samples were analyzed for carbon dioxide by gas chromatography with a thermal conductivity detector. Samples were analyzed at times ranging from 2 to 64 hours. The second method was to cut the plants back close to the soil, close the chamber and monitor the increase of CO_2 in the gas phase using the MIDAC system. In this case, spectra were taken at 10- or 15-min intervals for two hours. Then known amounts of CO_2, generated from K_2CO_3, were added and spectra recorded. This provided an internal calibration. A CO_2 band in the region of 2250 cm^{-1} and a line at 720 cm^{-1} were used for estimation of CO_2.

Molar absorptivities in the gas phase were determined for toluene, phenol and CO_2 by use of a gas cell of known pathlength. The same MIDAC instrument was used for this calibration as for the experimental measurements. For toluene and phenol, which are normally absent from the ambient air, a 5-cm pathlength cell was used. For CO_2, it was necessary to use a 50-cm cell because fluctuations of atmospheric CO_2 in the path between source, cell, and detector interfered with calibration runs in the short pathlength cell.

Results and Discussion

Flow regime and ground water toluene levels. Conditions varied slightly throughout the period of operation, but initially the inflow of water was 1 L per day when the growth chamber was not enclosed. After metal sides and a glass lid were added, the rate of water use decreased, presumably because the glass lid reduced the air circulation which was normally induced by operation of the hood exhaust fan. If the small fan within the chamber was not run, the surface of the soil was damp whereas when it was run, the surface remained relatively dry. Spider mites were an occasional problem and were controlled by spraying with malathion as needed. During very rainy weather, the water usage decreased noticeably. We have not attempted a precise correlation of water usage with relative humidity or plant growth stage.

Usually the chamber was operated with only minimal outflow of water, although for some experiments the rate was increased so that more than 0.5 L per day was exiting the tank. In Table I, some typical concentrations of toluene and phenol in the aqueous phase are shown for different days of similar water flow.

TABLE I Concentration of toluene and phenol in the saturated zone at several sampling wells

Substance	100 x Apparent absorbance in organic extract					
	Inlet	Port 1	Port 2	Port 3	Port 4	Outlet
toluene	38 ± 3	32 ± 2	29 ± 4	30 ± 2	27 ± 4	32 ± 1
phenol	76 ± 5	70 ± 7	56 ± 7	25 ± 3	5 ± 3	4 ± 2

Experiments were done on 5 consecutive days, with results shown as the mean ± standard deviation. Inlet flow was water saturated with toluene or having a fixed concentration of phenol. The inlet concentration of toluene was 515 mg/L while that of toluene was 0.5 mL/L of 93%, liquified, phenol.

The concentration of the organic chemical appears to stay the same (toluene) or decrease (phenol) as the water passes through the tank. If there were selective partitioning and exclusion of the compound at the surface of the plant root as described in the studies of Briggs et al. (*9*), one would expect the concentration to increase at the downstream end of the chamber. For toluene and phenol, the predicted partition coefficient, based on the K_{ow} near 100, is about 0.8. Thus if 2 L of solution enter the tank and 1 L is transpired, the concentration in the 1 L that exits as ground water should be elevated c. 20%. If the outflow is a smaller fraction of the input, say 1 L for every 5 L entering which is typical for the conditions that we used, the concentration of the compound in the output water should be considerably higher, approaching twice the input concentration. Preferential evaporation of water compared to phenol, if it occurred, would also have given an increase in phenol concentration in the ground water. The extent of increased concentration would depend on relative input and outflow rates and the extent of water loss by evaporation vs. transpiration. The relevant vapor pressures at room temperature are 25 mm for water and <1mm for phenol; thus preferential evaporation of water to concentrate the phenol is quite plausible.

For toluene, the concentration in the ground water stayed relatively constant throughout the course of flow through the chamber. This suggests that a degradation pathway for toluene was not induced in the saturated zone even after nearly 1 year of exposure. To verify that the UV-absorbing material observed in ground water was in fact toluene and not a degradation product, the entire UV spectrum of organic extracted material was determined. The spectrum could be directly superimposed on that of toluene, which has several characteristic features. Thus, the UV-absorbing material in the ground water remains as toluene.

On the other hand, the decrease in phenol concentration is quite striking, suggesting that either there is an anaerobic degradation system induced, or the ground water is being made aerobic during passage through the chamber. The extent of decrease in phenol concentration with distance of passage was much less obvious in results obtained two months earlier (*14*) than in the present results, indicating a time dependent induction of a degradative process. More recent studies using trichloroethylene and trichloroethane in place of phenol, have shown production of methane and chloride in the saturated zone of the latter half of the flow path, indicating that it is anaerobic.

Losses of input organic chemicals by adsorption to soil in this system are unlikely to be significant over the long term. Typical adsorption values for soil are in the range of 100 μg organic solute per g soil organic C (*16*). The chamber has been in operation for more than a year at high levels of input organic chemicals (c. 500 mg/L). Adsorption sites on the soil having only c. 1 % organic matter (c. 1500 g for the entire chamber) should be fully saturated within a few days of operation; adsorption is not likely to account for the continued disappearance of these organics. The simplest explanation is biodegradation in the soil.

The results shown in Table I were obtained with a mean water flow-through time of c. 4-7 days. This was estimated from the cross-sectional area below the water table available for flow (c. 1/3 of 10 x 10 cm), the length of the channel (1.8 m), the input rate of c. 1.0 L/day and output of < 400 mL/day. These estimates have been verified using KBr as a conservative tracer (unpublished). Mean flow-through time was usually greater than 7 days but could be made indefinitely long (i.e. no outflow) by controlling water input rate. From the water consumption rate (input minus output) of c. 600 mL/ day an average net input of >300 mg toluene per day was calculated. Toluene has a boiling point of 110°C and a vapor pressure of 30 mm at 25°C compared to water with a vapor pressure of 25 mm at 25°C. Toluene should not be significantly lost to the atmosphere by preferential volatilization in the capillary fringe and vadose zone, and it ought to be present in the vapor phase at about the same relative concentration as in the water phase, if it is not degraded.

Gas-phase toluene. Some examples of spectra obtained with this system are shown in Figures 1 & 2. The steady-state toluene concentration in the gas phase was below the level of detection (c. 250 ppb v/v) while the expected level based on water input rate was >50 ppm v/v per hour, in the gas phase. This is calculated using the input rate of 1 mL/min at 500 μL/L (c. 300 μmol/hr). This may be seen in Figure 1B, spectrum (17/18).

The primary IR absorption band of toluene is observed at 729 cm^{-1} as seen in Figure 1. A control experiment with toluene-saturated water applied to the surface of unadapted soil and immediately monitored in the same chamber accumulated gas-phase toluene levels of over 50 ppm v/v within two hours, as seen in spectrum (26/25). The enclosure was calibrated by direct release of known amounts of toluene into the gas phase (in theory 150 ppm v/v), as shown in spectrum (119/12). The prominent feature at 720 cm^{-1} is a CO_2 line. This line is an excited state, which is temperature dependent and so is probably not a good line for reliable monitoring of respiration rate. The plants were harvested prior to collecting spectrum (25/26) so that there was net accumulation of CO_2 in the chamber, whereas with the plants present and the lights on for spectra (17/18) and (119/12) the plants were actively photosynthesizing and depleted CO_2.

A second toluene band at 1033 cm^{-1} having about five-fold less intensity, was used to confirm the behavior of the system and to verify the identification of the toluene. Spectra shown in Figure 2 are the same as those of Figure 1. This latter spectral region was also monitored through windows of CaF_2. Windows of CaF_2 are much more water resistant than KBr, but energy transmission in the region of 730 cm^{-1} is insufficient for use with that spectral feature. We found the inconvenience of potential water damage to windows of less importance than the difference in sensitivity between spectral regions.

Similar gas phase studies were attempted with phenol. A calibration spectrum obtained in a 5-cm pathlength cell showed good detection of a concentration of less than 10^{-5} M (250 ppm v/v), indicating that with a 2.44-m path one would obtain good spectra with 5 ppm (v/v). The estimated limit of detection at 2.44-m pathlength was c. 0.15 ppm (v/v). If phenol volatilized at the same rate as water, the expected accumulation per hour would be at least 30 ppm (v/v).

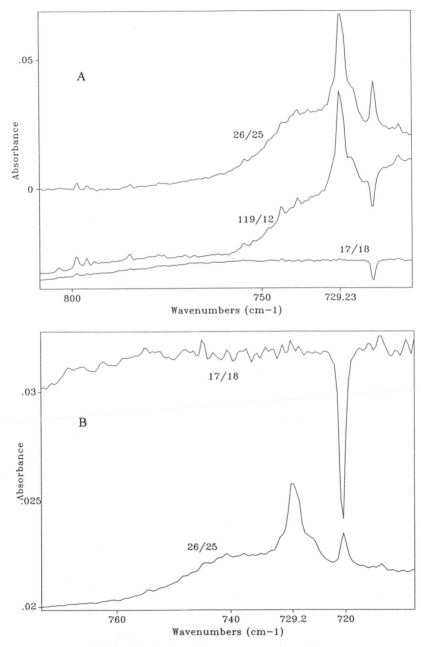

Figure 1. Main peak of toluene absorbance at 729 cm⁻¹. Part A: spectra of 15 hr equilibration vs open chamber (17/18); known amount of toluene (100 μL) evaporated in chamber (119/12); One L water saturated with toluene applied to surface of dry soil (26/25). All are displayed on same scale as indicated on left axis. Part B: (17/18) with the indicated scale; (26/25) at 10 fold less sensitivity.

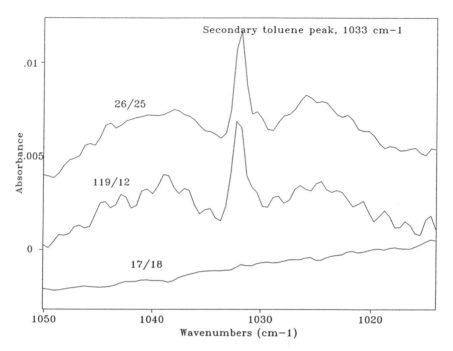

Figure 2. Secondary peak of toluene absorbance at 1033 cm[-1]. Spectra described in Figure 1A. Note the absorbance scale which indicates about five fold lesser intensity for this peak than that at 729 cm[-1].

Phenol concentrations were below the limit of detection, even when 2 mL of phenol was left in the chamber for 2 hours with the fan in operation. Thus, although 1 mm partial pressure of phenol in the gas phase would be easily detected, the rate of evaporation is too low for practical detection of phenol accumulation in a chamber that has a half-time for gas exchange on the order of 1-3 hours, or else the phenol adsorbed to hydrophobic surfaces rather than remaining in the gas phase. If phenol is taken up by the plants, it appears either to be volatilized much less rapidly than water, or to be metabolized to concentrations below the detection limit of the system. We prefer the latter explanation because if the phenol did not depart the leaves at a rate comparable to that of water, it would accumulate to toxic levels over the course of days. There was no evidence of toxicity to the plants (*14*).

More recently, experiments have been done with trichloroethane (TCA) and trichloroethylene (TCE) in the ground water. Accumulation of TCA, which is not easily biodegraded, is readily observable in the gas phase, under experimental conditions identical to those described here. Thus, a "conservative tracer" indicates that toluene would also be observable in the gas phase if it was not being degraded.

Respiration rate. Respiratory activity in the soil + rhizosphere was initially estimated from CO_2 evolution in the soil between plants as describe in Methods. The CO_2 levels detected tended to be higher near the inlet port than further along the flow channel, but the sum of all amounts, extrapolated to the total surface area of the chamber was only a few mmol/day, considerably less than the amount of toluene or phenol carbon entering the chamber. When accumulation of CO_2 was measured in the entire chamber by FTIR, much higher estimates were obtained.

This is presumably because a large fraction of total respiration occurs within, or in close proximity to the plants. Based on the initial rate of CO_2 accumulation and the internal calibration of the chamber, 70-100 mmol CO_2 was released per day. The input of dissolved organic C was 50 mmol/day or less. Turnover of photosynthetically fixed carbon in the rhizosphere, or via plant respiration, obviously contributes a significant fraction of the total respired CO_2. When CO_2 accumulation was monitored after withdrawal of toluene and phenol from the feed solutions, the apparent respiration rate decreased to about 2/3 of that previously observed, consistent with a significant contribution from the input organic chemicals.

For CO_2 accumulation measurement by FTIR, the plants were cut to 5 cm so that photosynthesis or dark respiration by the plant leaves and stems would not complicate the measurements. However, cutting may have stimulated root respiratory activity for initiation of regrowth (17). In addition, the plants are totally dependent on nitrogen (N) fixation for their source of nitrogen and this process requires a significant amount of respiratory activity. Plants accumulating 1 g dry matter per day need about 1/3 mmol N fixed per day. According to Herridge and Pate (18), this may require consumption of 8 mmol carbon per day by the root system for fixation of 1/3 mmol N, about 10% of the observed total respiration. Maxwell et al. (17) noted that roots may shut down nodular activity when the tops are removed so that the contribution of N fixation to respiratory demand may decrease upon cutting. They reported a general root respiratory rate of 0.9 mg CO_2/kg root matter/sec which may be converted to about 1.8 mmol/g root/day. If the mass of roots equals that of the harvested tops (47.5 g at one harvest), a respiratory rate of 70 mmol/day from a plot of 0.4 m^2 seems plausible. As an annual rate per hectare, it is also consistent with reports of soil respiration world-wide as cited by Glinski and Stepniewski (19). They cite values for a range of crops and climates from very small up to 28 L/day/m^2 while our estimate is about 4.5 L/day/m^2, near the median of the range quoted by Glinski and Stepniewski.

Conclusions

The work described here shows that with an adapted system plants are able to grow actively for one year in the presence of water saturated with toluene, or containing c. 500 ppm phenol. There is little intermedia transfer of toluene, presumably because there is little uptake by the plants or direct evaporation from the soil. Total system respiratory activity, measured as carbon dioxide evolution, is more than adequate to account for complete mineralization of the input toluene. Only a small portion of the toluene leaves the system, in the exiting ground water. Phenol, a less volatile compound which could not be detected in the gas phase, in part because of its low volatility, appears to be effectively degraded also. None of it leaves in the exiting ground water.

Acknowledgments

This research was partially supported by the US EPA under assistance agreements R-815709 and R-819653 to the Great Plains-Rocky Mountain Hazardous Substance Research Center for regions 7 and 8 under project 90-13 and an EPA grant (CR81-7790-01-1) to Drs. Fateley and Hammaker. It has not been submitted to the EPA for peer review and therefore may not necessarily reflect views of the agency and no official endorsement should be inferred. The US Department of Energy, Office of Restoration and Waste Management, Office of Technology Development and the Center for Hazardous Substance Research also provided partial funding. This is contribution number 94-253-B of the Kansas State Agricultural Experiment Station.

Literature Cited

1. Aprill, W.; Sims, R.C. *Chemosphere* **1990** *20*, 253-265
2. Walton, B.T.; Anderson, T. A. *Appl. Environ. Microbiol.* **1990** *56*, 1012-1016
3. Schnoor, J.L.; Licht, L. A. **1991** In *Conference on Hazardous Waste Research*; Kansas State University; Manhattan, KS
4. Shimp, J.F.; Tracy, J.C.; Davis, L.C. Lee, E.; Huang, W.; Erickson, L.E. (1993) *Crit. Rev. Environ. Sci. Tech.* **1993** *13*, 41-77
5. Davis, L.C.; Erickson, L.E.; Lee, E.; Shimp, J.; Tracy, J.C. *Environ. Progress* **1993** *12*, 67-75
6. Tracy, J.C.; Erickson, L.E.; Davis, L.C. **1993** *Proceedings Air Waste Management Assoc. Ann. Meeting*
7. McFarlane, J.C.; Cross, A.; Frank, C.; Rogers, R.D. (1981) *Environ. Monit. Assess.* **1981** *1*, 75-81
8. National Research Council. **1993** In Situ Bioremediation. When Does it Work? National Academy Press, Washington, D.C. 20418
9. Briggs, G.G.; Bromilow, R.H.; Evans, A.A. (1982) *Pestic. Sci.* **1982** *13*, 495-504
10. McFarlane, J.C.; Pfleeger, T.; Fletcher, J. *J. Environ. Qual.* **1987** *16*, 372-376
11. Boersma, L.; Lindstrom, F.T.; McFarlane, C. Oregon State University Agric. Exp. Sta. Bulletin 677, **1990,** Model for uptake of organic chemicals by plants
12. Arvin, E.; Jensen, B.K.; Gundersen, A.T.; Mortensen,E. In *Organic Micropollutants in the Aquatic Environment*; Angeletti, G. and Bjorseth, A.; Kluwer Academic Publishers: Dordrecht, **1991**; pp 174-183.
13. Aharonson, N.; Katan, J. (1993) *Arch. Insect. Biochem.* **1993,** *22*,451-466
14. Davis, L.C.; Chaffin, C.; Muralidharan, N.; Visser, V.P.; Fateley, W.G.; Erickson, L.E.; Hammaker, R.M.; In *Conference on Hazardous Waste Research*; Kansas State University, Manhattan, KS, 1993;
15. Wacek, T.J.; Brill, W.J. *Crop Sci.* **1976** *16*, 519-522
16. Fan, S.-F.; Scow, K.M. *Appl. Environ. Microbiol.* **1993** *59*, 1911-1918
17. Maxwell, C.A.; Vance, C.P.; Heichel, G.H.; Stade, S. *Crop Sci.* **1984** *24*, 257-264
18. Herridge, D.F.; Pate, J.S. *Plant Physiol.* **1977** *60*, 759-764
19. Glinski, J.; Stepniewski,W. *Soil Aeration and Its Role for Plants;* CRC Press: Boca Raton, FL, 1985

RECEIVED February 15, 1994

Chapter 11

Volatilization and Mineralization of Naphthalene in Soil–Grass Microcosms

J. W. Watkins, D. L. Sorensen, and R. C. Sims

Civil and Environmental Engineering Department, Utah State University, Logan, UT 84322–4110

The potential for vegetation-enhanced biodegradation of naphthalene in artificially contaminated soil was studied in laboratory microcosms. Microcosms containing soil without plants and soil supporting two-month old Bell Rhodesgrass were treated with naphthalene and spiked with [7-^{14}C]naphthalene. Compressed air was continuously passed through each microcosm, through a trap to collect volatile organics, and through a trap for CO_2. The microcosms were incubated under artificial lighting with a 16 h photoperiod for 25 days. After incubation, soil was solvent extracted and combusted to recover bound radiolabel. Volatilization losses during operation and analysis prevented reaching a mass balance of the radiolabel. Naphthalene volatilization was enhanced by vegetation but mineralization was decreased in vegetated microcosms in comparison to those without vegetation.

Polycyclic aromatic hydrocarbon (PAH) compounds may be toxic and carcinogenic. Soils can become contaminated with PAH compounds from many industrial sources including wood preservatives, coal gasification wastes, and petrochemical wastes. Many PAH compounds, especially those with two and three ring structures, are biodegradable and can serve as growth substrate for microorganisms. Higher molecular weight PAH compounds may be cometabolized by soil microorganisms, with soil half-lives of several hundred days (1). Plant-enhanced biodegradation of PAH compounds is a recent area of research and is based on the hypothesis that increased microbial activity associated with plant roots will accelerate biodegradation. Aprill and Sims (2) showed a statistically significant increase in disappearance of benz(a)anthracene, benzo(a)pyrene, chrysene and dibenz(a, h)anthracene from soils vegetated with eight different prairie grasses compared to unvegetated soils. Wheat straw (3), trichloroethylene (4), surfactants (5), parathion (6, 7), and diazinon (7) have all been shown to degrade faster in soil/plant systems when compared with soil degradation alone.

Naphthalene is the smallest PAH and has the physiochemical properties listed in Table I. The partition coefficients in Table I were calculated from structure-

0097–6156/94/0563–0123$08.00/0

activity relationships (8). Naphthalene was biodegraded in soil relatively rapidly under aerobic conditions (9) and may be distributed among the gas, liquid, and solid phases in unsaturated soil (10, 11, 12, 13, 14). Environmental fate models that use partition coefficients sometimes exclude the affects of biota on the fate of chemicals (15). The predominant mechanism(s) (i.e., sorptive immobilization, translocation by leaching or volatilization, or biodegradation) effecting the fate of naphthalene in contaminated soil

Table I.   Physical properties and degradation half-life of naphthalene

Property	Value	Reference
Molecular Weight	128	
Aqueous solubility	30 mg/L	(1)
Vapor Pressure (20° C)	0.049 torr	(8)
log K_d	0.67	(8)
log K_{oc}	3.11	(8)
log K_{ow}	3.37	(8)
log K_h	-1.97	(8)
$t_{1/2}$	2.1 days	(8)

K_d = concentration in soil/concentration in water.
K_{oc} = K_d/percent soil organic carbon,
K_{ow} = concentration in octanol/concentration in water.
K_h = concentration in air/concentration in water. (Henry's law constant).
$t_{1/2}$ = Half-life; time required for degradation of 1/2 of the original concentration in soil.

will depend on the physical, chemical, and biological environment of the contaminated soil, the soil below, and the atmosphere above the contaminated zone. Due to the high vapor pressure of the naphthalene, as compared to other PAH compounds, significant volatilization losses may be expected under most soil conditions.

The present study sought to determine the affect of vegetation on the fate of naphthalene in soil microcosms with emphasis on the mechanisms of volatilization and biological mineralization.

Materials and Methods

Triplicate microcosms with three different treatments were used. First, microcosms containing soil poisoned with $HgCl_2$, second, microcosms without vegetation, and third, microcosms with vegetation. Soil used in all of the microcosms was Kidman sandy loam (Typic Haplustoll) collected at a Utah Agricultural Experiment Station farm at Kaysville, Utah (Table II). The soil had no history of pesticide application. Each microcosm consisted of a glass, 250 mL Erlenmeyer wide-mouth flask fitted with a two-holed neoprene stopper (Figure 1). Each flask was wrapped with black plastic tape from the bottom up to about 0.3 cm above the soil surface to discourage algae growth. The stopper was fitted with two disposable glass Pasteur pipettes and an 18 gage, 3.8 cm syringe needle. The pipettes were used for gas exchange and the needle for watering the soil. A one-way, male Luer-Lok stopcock valve was attached to one of the pipettes with clear silicone adhesive to control incoming air. The adhesive was allowed to cure for three days prior to use. The air inlet pipette was positioned so that the narrow end of the pipette was about 1 cm above the soil surface. The discharge pipette extended below the stopper about 0.3 cm. A fourth microcosm for each treatment was also prepared, without naphthalene addition, as a control.

Traps for volatile organic compounds (VOC) and CO_2 in discharge air were arranged in tandem for each microcosm. Each trap consisted of a 25 X 150 mm borosilicate glass culture tube closed with a two-holed neoprene stopper fitted with two Pasteur pipettes arranged so the incoming air was carried to the bottom of the trap and discharge air exited near the top. Neoprene tubing connected the microcosm to an air supply manifold and the traps. The first trap was filled with 40 mL ethylene glycol monomethyl ether (J. T. Baker, 99+%) to collect VOC. The second trap was filled with 40 mL of Ready Gel scintillation counting liquid (Beckman), methanol, and monoethanolamine (J. T. Baker) in a 5:4:1 mixture, respectively. Abbott et al. (*16*) reported trapping efficiencies in excess of 98% for ^{14}C-naphthalene and [14]CO_2 in similar traps.

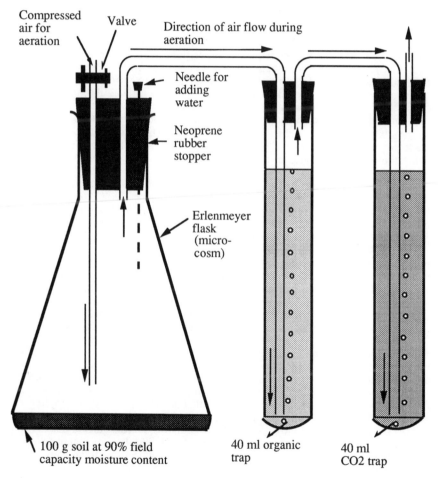

Figure 1. Schematic drawing of the microcosm apparatus used in the experiment.

Table II. Physiochemical properties of Kidman fine sandy loam

Organic carbon (%)	0.68
pH	7.8
Electrical Conductivity (mmhos/cm)	0.5
Cation Exchange Capacity (meq/100 g)	12.1
Percent moisture at -33 kPa	12.4
Percent moisture at -1500 kPa	5.5
NO_3-N (mg/kg)	4.7

Each microcosm in the abiotic and unvegetated treatments contained 100 g of air dried soil which was watered to 90% of field capacity (11% moisture). Each vegetated microcosm was loaded with 75 g of soil, planted with 1 g of *Cloris gayana* seed, followed by 25 g of soil. All of the microcosms were watered to 90% of field capacity, closed with paraffin film, and allowed to stand for 24 hours to permit moisture equilibration through the soil. Each of the poisoned microcosms was watered with a 2% $HgCl_2$ solution (*16*).

The microcosms were incubated in a ventilated, air conditioned room under a 1000 watt metal halide light (Venture) and two banks of fluorescent light fixtures each with four, 40 watt, 1.22 m plant lights (General Electric, Plant Light; Sylvania, Gro Lux) with a 16 hour photoperiod. Average light and dark temperatures were 27° and 16° C, respectively. Moisture content was maintained between 60 and 90% of field capacity (7 and 11% moisture) throughout the experiment by periodically adding water through a hypodermic needle temporarily inserted through the stopper (Figure 1). The microcosms were randomly moved to new positions every seven days throughout the experiment. Plants were allowed to grow in the microcosms for 60 days before being treated with ^{14}C-naphthalene. This allowed the above ground plant to grow to the neck of the flask and roots had grown throughout the soil and along the bottom and sides of the glass. All of the plant was retained within the microcosm.

Each microcosm was loaded with 0.5 mL of a solution of naphthalene dissolved in methanol. Sufficient unlabeled crystalline naphthalene (J. T. Baker, 99+%) and [7-^{14}C]naphthalene (Sigma, 98+%) was dissolved in methanol so that each microcosm receiving the radiolabel could be spiked with 10 mg unlabeled naphthalene and 0.0032 mg of ^{14}C-naphthalene. The spiking solution averaged (n=3) 576,000 disintegrations per minute (DPM). Ten uniformly spaced holes were made in the microcosm soil with a 0.6 cm diameter glass rod and 50 μL of the naphthalene solution was injected into the bottom of each hole. A drop of deionized water was immediately added to each hole to help disburse the naphthalene. This technique minimized the amount of solvent used to deliver the naphthalene and minimized the apparent toxicity (leaf yellowing) to the plants while allowing the desired concentration of naphthalene to be added and dispersed in the soil.

Each microcosm and its traps were stoppered and connected to an air manifold. Breathing quality compressed air, supplied at 4 psi pressure, was passed through each microcosm at a rate of 25 mL per minute by adjusting the stopcock valve on the air inlet pipette. At each sampling, the air supply pressure was reduced to zero and each trap was disconnected, emptied, and rinsed with several ~5 mL volumes of methanol into four 20 mL glass liquid scintillation (LS) vials. Total sampling time for all microcosms was approximately 2 hours.

At the termination of the experiment, each microcosm was quickly disconnected from the air manifold, placed on ice and flooded with 50 mL of methanol. Stoppers that sealed each microcosm and set of traps and all connecting tubing were quickly disconnected and rinsed several times with methanol back into the microcosms. The microcosms were sealed with paraffin film. Above ground plant tissue was cut off at the soil surface and sealed in glass culture tubes with paraffin film. The microcosms and tubes with plant tissue were then stored in the dark at -20° C until analyses were performed.

The soil, root-soil complex, and plant tissue were extracted with chilled methanol. Methanol was chosen as an extractant because it is miscible with water and saturates the soil without causing clumping. Fifty milliliters of methanol was added to an empty flask and analyzed with each batch of analyses as a method blank. The soil and methanol suspension in each flask was blended for 3 minutes at 30% of full speed with a tissumizer (Tekmar). The soil was allowed to settle from the suspension and the supernatant was vacuum filtered through Whatman No. 4 filter paper in a Buechner funnel. An additional 50 mL of chilled methanol was added to the slurry and the extraction was repeated. The slurry was transferred, with rinsing (3X, 20 mL), to the Buechner funnel and the filtrate was combined with the first extract. The extracted soil was returned to the microcosm flask which was sealed with paraffin film. The pooled filtrate was dried by passing it through a Na_2SO_4 column and concentrated to 10 mL in Kuderna-Danish glassware by heating in a 90° C water bath. The concentrate was transferred, with rinsing, to a 20 mL scintillation vial containing 5 ml of Ready Gel (Beckman) scintillation cocktail.

Naphthalene volatilization potential was evaluated using method controls, containing only methanol and radiolabeled naphthalene, which were carried through all steps of analyses. An additional triplicate set of method controls were extracted without blending and filtration. A check for leaks in the microcosm-trap arrangement was done using methanol and radiolabeled naphthalene in triplicate microcosms without soil. Air was passed through these microcosms at 25 mL/min for 24 hours. The contents of the microcosms were then rinsed into scintillation vials and the VOC traps were processed as described above. Naphthalene volatilization from non-vegetated soil microcosms was evaluated by adding 100 g of air dried Kidman soil to nine microcosms and adding water to bring the soil to 90% field capacity. A single 0.3 cm hole was made in the soil at the center of the microcosm and 0.5 mL of methanol containing radiolabeled naphthalene was placed in the hole. A liquid scintillation vial containing 15 mL of VOC trapping solution (EGME) was suspended inside each microcosm from the stopper used to seal the microcosm. The microcosms were incubated under the same conditions used in the vegetation experiment and three microcosms were removed for analysis after 0.25, 4, and 24 hours. The VOC trap scintillation vials were filled with Ready Gel scintillation fluid, capped, and counted.

Above ground plant tissue was cut into 1.3 cm lengths and ground to powder in liquid nitrogen in a mortar and pestle. The ground tissue was rinsed into an Erlenmeyer flask containing 50 mL of chilled methanol and allowed to sit in an ice bath for 24 hours before blending and concentrating in the manner described above for soil. Filtering of plant extracts was by gravity. A single, 0.5 g plant tissue sample was analyzed for moisture content.

Computerized statistical analysis of variance (ANOVA) procedures (JMP, SAS Institute) were used to determine the significance of differences among treatments. A repeated measures ANOVA design was used where measurements were made from the same microcosms over time.

Results and Discussion

A summary of the distribution of radiolabeled carbon in the various microcosms is shown in Table III. Unfortunately, there were losses of more than 50% of the naphthalene and/or its degradation products during operation and analysis of the microcosms. This prevented reaching a mass balance of the radiolabel added to the microcosms. Method controls, containing only methanol and radiolabeled

Table III.   Percent distribution of radiolabeled carbon in the microcosms. The mean ± standard deviation are shown

Microcosm	VOC	CO_2	Percent DPM Extract	Oxidation	Total[†]
Poisoned	16 ± 3	0	2 ± 0.1	1 ± 0.3	19 ± 3
Unvegetated	10 ± 2	18 ± 8	2 ± 0.8	19 ± 7.3	49 ± 11
Vegetated	15 ± 4	5 ± 1	1 ± 0.7	8 ± 1.7	NA
Plant tissue	NA	NA	0 ± 0.1	1 ± 0.4	30 ± 5

NA = Not applicable
[†]Sum of VOC, CO_2, Extract, and Oxidation for the microcosm soil and plant tissue.

naphthalene, which were carried through all steps of analyses, lost 23% of the radiolabel during tissue extraction, and 25% of the radiolabel during soil/root complex extraction. The method controls extracted without blending and filtration lost 10% of the radiolabel. An average of 45% of the radiolabel in the microcosms with methanol and naphthalene that were flushed with 25 mL air per minute for 24 hours was not recovered and was apparently lost to leaks. The average percent radiolabel collected in the scintillation vials suspended in the static headspace of the soil microcosm was 0.21, 2.3, and 8% for 15 minutes, 4 hours, and 24 hours, respectively. Assuming simple first order kinetics for volatilization in this case, linear regression of the natural logarithm of the average amount of label remaining with time yields a kinetic constant, k, of 3.28 X 10^{-3}/h and, hence, a volatilization half-life of about 9 days. This potential for relatively rapid volatilization from soil combined with the leaks in the gas scrubbing systems and the potential loss of naphthalene during analysis, indicates that volatilization losses were the primary cause for not achieving a mass balance in the experiment.

The variations in the losses of radiolabel appeared to be randomly distributed among all of the microcosms used in the vegetation experiment and were considered to be a component of the error in the VOC and $^{14}CO_2$ measurements from the traps. With this assumption, it is possible to compare the volatilization and mineralization of naphthalene in the vegetated microcosm experiments.

The measured volatilization of radiolabeled compounds over the 24 day experiment is illustrated in Figure 2. Volatilization was significantly higher in the microcosm with plants than in the soil microcosms from the first day of the experiment (i.e., the interaction of "treatment" and sampling "day" were significant; $p \leq 0.05$). The rate of volatilization was apparently higher in the vegetated microcosms for about the first 5 days but was similar to that in the soil microcosms for the remainder of the experiment. These data suggest that naphthalene was taken up by the grass roots, translocated within the plant, and volatilized through the above ground plant structures. This mechanism of removal of naphthalene from the rhizosphere environment would be expected to reduce the amount of naphthalene available for biodegradation. If the volatilization of naphthalene from contaminated soil is a common phenomenon, the establishment of vegetation as a treatment measure may have air quality monitoring and regulatory compliance implications.

Carbon dioxide evolution from the microcosms is summarized in Figure 3. Again, the interaction of "treatment" and sampling "day" was significant. The difference in mineralization of naphthalene was not significant until after 11 days ofincubation.   The rate of mineralization in the soil microcosms increased substantially after 9 days of incubation while the mineralization rate in the vegetated microcosms slowed slightly after 9 days. This may be the period required for adaptation of the soil microbial community to be able to utilize naphthalene. It is noteworthy that voltilization from both the soil and vegetated microcosms slowed

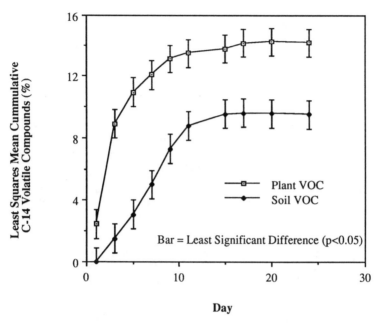

Figure 2. Cumulative volatilization of radiolabeled organic compounds in vegetated and unvegetated microcosms.

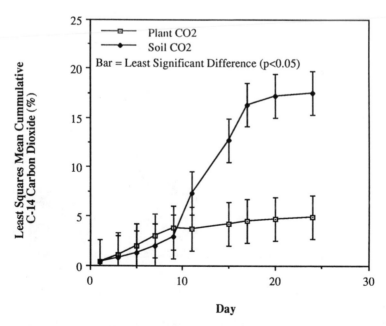

Figure 3. Cumulative $^{14}CO_2$ trapped from vegetated and unvegetated microcosms.

during the 7 to 11 day period of the experiment. This suggests that there may have been an interaction between volatilization and mineralization activity. It may be that volatilization removed naphthalene, reduced toxicity and thus contributed to the onset of mineralization activity. Mineralization, in turn, removed naphthalene and reduced the volatilization potential.

The reason for the relatively low mineralization of naphthalene in the vegetated microcosms is unknown. Possible contributing factors may include reduced availability of naphthalene through plant assimilation and volatilization. Less than 3% of the C-14 was recovered from above ground plant tissue (Table III) but it is possible that some of the $^{14}CO_2$ produced through mineralization was incorporated into the plant through photosynthesis. In addition, the increased availability of labile organic matter from the plant roots may have placed naphthalene degrading microorganisms at a disadvantage in the competition for carbon and energy substrate. Those organisms with genotypes for naphthalene degradation may not have induced enzymes for naphthalene utilization in the presence of other readily available substrates.

The average amounts of C-14 bound to soil (not solvent extractable) at the end of the experiment are shown graphically in Figure 4. Significantly more C-14 was bound to the soil in the unvegetated microcosms than in the poisoned microcosms. The average amount of C-14 bound to the vegetated microcosm soil was lower than that in the unvegetated microcosms but it was not statistically different from the unvegetated or poisoned microcosm soil. Three percent or less of the added C-14 was solvent extractable from the soil (Table III). These data support the hypothesis that microbial activity is involved in binding naphthalene or its degradation products to the soil.

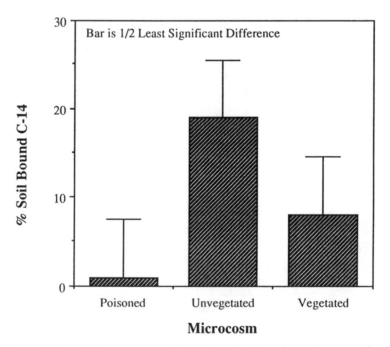

Figure 4. The relative amounts of C-14 bound to the poisoned, vegetated, and unvegetated soil in the microcosms.

Conclusions

Based on the assumption that losses of C-14 through volatilization during analysis procedures and leakage of volatiles and CO_2 during microcosm operation are random, it is possible to conclude that Bell Rhodesgrass enhanced volatilization of naphthalene from the fine sandy loam soil used in the microcosms. Mineralization of naphthalene was suppressed by the grass in comparison to that in unvegetated soil. More information is needed on the volatilization and mineralization affects of various plants growing in different soils before any conclusions should be drawn about the use of vegetation to enhance the removal of PAH contaminants from soil. Other research, cited above, has concluded that soil PAH removal is enhanced by the presence of vegetation. Although the role of the rhizosphere microbial community on the humification (soil incorporation) of C-14 is unclear from the data obtained in the present study, it is clear that microbial activity is important in this process.

Literature Cited

1. Sims, R. C.; Overcash, M. R. *Residue Rev* . **1983**, *88*, 1-21.
2. Aprill, W.; Sims, R. C. *Chemosphere*, **1990**, *20*, 256-265.
3. Cheng, W.; Coleman, D. C. *Soil Biol. Biochem*, **1990**, *22*, 781-787.
4. Walton, B. T.; Anderson, T. A. *Appl. Environ. Microbiol.*, **1990**, *56*, 1012-1016.
5. Federle, T. W.; Schwab, B. S. *Appl. Environ. Microbiol.*, **1989**, *55*, 2092-2094.
6. Reddy, B. R.; Sethunathan, N. *Appl. Environ. Microbiol.*, **1983**, *45* , 826-829.
7. Hsu, T. S.; Bartha, R. *Appl. Environ. Microbiol.*, **1979**, *37* , 36-41.
8. Sims, R. C.; Doucette, W. J.; McLean, J. E.; Grenney, W. J.; Dupont, R. R. *Treatment potential of 56 EPA listed hazardous chemicals in soil*. EPA/600/S6-88/001. U. S. Environmental Protection Agency, Cincinnati, OH.
9. Bossert, I. B.; Bartha, R. *Bull. Environ. Contam. Tox.*, **1986**, *37*, 490-495.
10. Sims, J. L.; Sims, R. C.; Matthews, J. E. *Haz. Waste Haz. Mat.*, **1990**, *7*, 117-149.
11. Koskinen, W. C.; Harper, S. S. In *Pesticides in the soil environment: processes, impacts, and modeling;* Cheng, H. H., Ed.; Soil Science Society of America, Inc., Madison, WI, 1990; pp 51-77.
12. Hutzler, N. J.; Gierke, J. S.; Krause, L. C. In *Reactions and movement of organic chemicals in soils;* Sawhney, B. L.; Brown, K., Eds.; SSSA Special Publication 22; Soil Science Society of America, Inc., Madison, WI, 1991; pp 373-403.
13. Valentine, R. L.; Schnoor, J. L. In Vadose zone modeling or organic pollutants; Hern, S. C.; Melancon, S. M., Eds.; Lewis Publishers, Chelsea, MI, 1988; pp 191-222.
14. Huddleston, R. L.; Bleckmann, C. A; Wolfe, J. R. In *Land treatment: a hazardous waste management alternative;* Loehr, R. C.; Malina; J. F., Jr., Eds.; Water Resources Symposium 13; Center for Research in Water Resources, University of Texas, Austin, TX, 1986; pp 43-61.
15. Fletcher, J. S.; MacFarland, J. E.; Pfleeger, T; Wickliff, C. *Chemosphere*, **1990**, *20*, 513-523.
16. Abbott, C. K.; Sorensen, D. L.; Sims, R. C. *Environ. Toxicol. Chem.*, **1992**, *11*, 181-185.
17. Wolf, D. C.; Dao, T. H.; Scott, H. D.; Lavy, T. L. *J. Environ. Qual.*, **1989**, *18*, 39-44.

RECEIVED February 18, 1994

Chapter 12

Biologically Mediated Dissipation of Polyaromatic Hydrocarbons in the Root Zone

A. P. Schwab[1] and M. K. Banks[2]

Departments of [1]Agronomy and [2]Civil Engineering,
Kansas State University, Manhattan, KS 66506

Soil contaminated with petroleum sludge is often bioremediated by tillage and addition of nitrogen and phosphorus to stimulate microbial degradation of the hazardous compounds. This "landfarming" technique is effective in reducing concentrations of a variety of chemicals during the early stages of treatment, but degradation rates severely decline thereafter, especially for recalcitrant compounds such as polyaromatic hydrocarbons (PAHs). However, the presence of vegetation can enhance the degradation of these compounds. In a greenhouse experiment, the degradation of PAHs was greater in the presence of plants than in their absence. Target PAHs were detectable in the plant tissue, but the total quantity of uptake was insignificant. Enhanced microbial activity was apparently responsible for increased dissipation of target PAHs. Establishment of vegetation holds promise as an inexpensive yet effective means of passive remediation of sites contaminated with petroleum hydrocarbons.

A common environmental problem associated with pumping and refining crude oil is the disposal of petroleum sludge. The sludge is usually very viscous or solid and can be treated by incorporation into soil. If the soil is frequently tilled and fertilized, soil microorganisms will be stimulated and organic contaminants biodegraded. Unfortunately, biodegradation of the more recalcitrant and potentially toxic contaminants, such as the polyaromatic hydrocarbons (PAHs), is rapid at first but quickly declines.

The establishment of vegetation in soils contaminated with hazardous organics can help stimulate bioremediation (1). Generally, plants will grow well only in moderately contaminated soil, and the use of plants to enhance biodegradation of organics is limited to the upper 2 m of the soil profile (2).

0097–6156/94/0563–0132$08.00/0

Despite these restrictions, establishment of vegetation can be an economic, effective, low maintenance approach to waste remediation and stabilization. Plants can interact directly with hazardous organic compounds through accumulation or degradation (3). Uptake of the contaminant by plant roots can be an active process involving specific enzymes and the expenditure of energy or a passive process in which neutral, hydrophilic compounds move unimpeded into the roots along with water (4). In either case, uptake of the pollutant is usually a direct function of concentration in the soil solution. Within the plant, the contaminant can be irreversibly adsorbed (2), accumulated in cells, or transported in the plant's vascular system (4). Plants often are able to metabolize toxic organics, and plant species tolerant to specific compounds can possess metabolic protection and significantly degrade an accumulated chemical (5).

The content of PAHs in plants has been assessed in many urban, agricultural, and pristine environments. Concentrations in leaves generally range from <1 up to 100 μg/kg (6). A number of studies at different locations (7,8) strongly indicate that atmospheric deposition is the primary pathway for accumulation of PAHs by plant leaves (9,10). Research focusing on the uptake of PAHs (benzo[a]pyrene, anthracene, and others) from contaminated soil (11,12) found that translocation of PAHs with four or more rings from roots to shoots was negligible. Increased plant concentrations were observed for PAHs with two or three rings, but there was no correlation between uptake and soil levels (13,14). Elevated fluoranthene, pyrene, and benzo[a]pyrene concentrations in leaves and stems were chiefly derived from atmospheric inputs, and root PAH content was probably a result of adsorption to the root surface.

Plants can contribute indirectly to the dissipation of contaminants in vegetated soil through a number of mechanisms. Increased microbial numbers and populations are often encountered in the soil immediately adjacent to the root (rhizosphere soil) (15). Plant roots are known to secrete sugars, amino acids, organic acids, and nucleic acid derivatives (16), and these compounds can serve as sources of carbon, nitrogen, and energy for the microflora. Microbial activity in the root zone stimulates root exudation that helps maintain high populations and elevated levels of metabolism.

Because roots permeate throughout the upper 30 cm of the soil profile and often have very high surface areas, adsorption onto the root surface is a potentially important sink for hazardous organic compounds. A study of the adsorption of pesticides onto the roots of several agricultural plant species demonstrated that marked differences in adsorption capacity occurred among plant species and contaminants (17). Adsorption also can be affected by the stage of plant growth due to changes with age in root structure and properties of the cell membrane.

The majority of the research in the field of biodegradation of organics in the rhizosphere has been limited to pesticides, with some work on trichloroethylene (TCE) and PAHs. Rhizosphere organisms responsible for the degradation of specific pesticides have been isolated (18) and enumerated (19).

Pesticide degraders were as much as 100 times more abundant in rhizosphere soil than in unvegetated soil (20). Degradation of [^{14}C]TCE in rhizosphere soil after the roots had been removed was 15% greater than in non-rhizosphere soil (20). Similar results were observed in systems with the roots intact. Aprill and Sims (21) studied the effects of using deep-rooted grasses to remediate soil contaminated with the PAHs benzo[a]pyrene, benzo[a]anthracene, chrysene, and dibenzo[a,h,]anthracene. The fibrous nature of the roots resulted in very high surface areas and active microbial populations that maximized bioremediation. After 219 days of plant growth, PAH dissipation was greater in the rhizosphere soil than non-rhizosphere soil, and the extent of degradation was directly proportional to water solubility of the targeted PAH.

Experimental Design.

Soils contaminated with petroleum hydrocarbons are fairly common throughout the United States. These sites are frequently found in areas of active oil pumping and processing. In cooperation with Conoco, Inc. of Ponca City, Oklahoma, a greenhouse research project was initiated to investigate the use of vegetation to reduce the concentrations of PAHs in landfarmed soil.

Contaminated and uncontaminated soil was collected from a disposal site near Davis, Oklahoma. The soil was mapped as a Lela clay. The Lela clays are Fine, mixed, thermic Udic Chromusterts. Three treatments were established: uncontaminated soil, uncontaminated soil spiked with PAHs, and contaminated soil spiked with PAHs. Several PAHs, including pyrene and anthracene, were added by dissolving in acetone and spraying onto the soil to achieve a concentration of 100 mg/kg for each compound. Anthracene and pyrene were chosen because they are found in petroleum-contaminated soils and differ in their resistance to degradation. One kilogram of soil was packed into a greenhouse pot and watered to field capacity. Pots were either left without plants or planted with tall fescue (*Festuca arundinacea*), sudan grass (*Sorghum vulgare*), switch grass (*Panicum virgatum*), or alfalfa (*Medicago sativa*). Plants were watered as needed; fertilized every 14 days; and harvested after 4, 8, 16, and 24 weeks. At harvest, plants were separated into shoots and roots, and the soil was collected from each pot. Each combination of soil and plant species was replicated four times for each time period.

Plants were analyzed for PAHs by solvent extraction and gas chromatography (22). Soils were analyzed by solvent extraction, reverse phase chromatography for pre-enrichment, and gas chromatography (23). Extraction efficiencies for PAHs from soils and plant tissue were greater than 92%. Soil from the pots growing alfalfa was used to enumerate microbial numbers by plate counts (24). Soil from alfalfa pots also was used to study the evolution of $^{14}CO_2$ from [^{14}C]pyrene in microcosms containing 20 g of soil. Treatments included rhizosphere soil (no roots), rhizosphere soil amended with organic acids, non-rhizosphere soil, and sterilized non-rhizosphere soil (24). Adsorption of naphthalene onto alfalfa and fescue roots was examined by equilibrating excised roots with solutions of different naphthalene concentrations and measuring the adsorbed and equilibrium solution concentrations (23).

Results

Plant Biomass. A prerequisite for the use of vegetative remediation is that the chosen plants must grow well in the contaminated medium. An excellent indication of plant health is total accumulated dry matter, or biomass. In this study, plant biomass followed a normal pattern of increase with time for all soils. After 24 weeks, the biomass of all species was well within the range expected for healthy plants (Table I). An initial concern existed that the contaminated soil was unsuitable for plant growth because soluble salts and sodium in the contaminated soil (2500 ppm salts and 4.8% exchangeable Na) were higher than in the uncontaminated soil (1200 ppm salts and 0.6% exchangeable Na). However, these plants apparently can survive in the presence of PAHs and other petroleum contaminants.

Table I. Biomass of roots and shoots after growing for 24 weeks in test soils

soil treatment	fescue		alfalfa		sudan grass		switch grass	
	shoot	root	shoot	root	shoot	root	shoot	root
	------------------------------------- g -------------------------------------							
uncontaminated soil	11.3	3.3	4.2	2.2	20.1	4.8	9.3	7.2
uncontaminated+PAH	7.1	1.7	5.0	2.4	17.0	4.6	17.6	13.0
contaminated+PAH	6.6	2.9	2.3	1.0	7.9	1.6	6.2	5.5
l.s.d. (P<0.05)	11.2	5.1	2.2	2.0	12.5	2.9	7.4	6.8

Plant Uptake. The mean uptake of pyrene per pot (Table II) for all plant species ranged from 0 to 30 μg ($\leq 0.03\%$ of total applied). Many of the detections of pyrene were in plants growing in uncontaminated, unspiked soil. This may be explained by either atmospheric deposition of hydrocarbons on the leaves. Surface deposition has been described previously as the predominant means of accumulation of PAHs in the above-ground portions of plants (9,10). This is possibly due to uptake from vapor in surrounding air or external contamination of shoots by soil and dust followed by retention in the cuticle. Compared to the total mass of PAHs applied to the soils, the PAH content of the roots and shoots was negligible.

Adsorption of Naphthalene onto Plant Roots. In the adsorption experiment, mature roots of alfalfa and fescue (22 weeks old) were equilibrated for 12 hours with solutions of naphthalene ranging in concentration from 0.5 to 5 mg/L. Equilibrium concentrations in solution ranged from 0.03 to 1.1 mg/L, and

Table II. Uptake of pyrene by alfalfa and fescue roots and shoots after 24 weeks of plant growth

compound	soil treatment	alfalfa		fescue	
		shoot	root	shoot	root
		------------------- μg/pot ----------------			
anthracene	uncontaminated	0	0	0	0
	uncontaminated+PAH	0	0	10	10
	contaminated+PAH	0	10	0	0
pyrene	uncontaminated	10	10	20	30
	uncontaminated+PAH	10	0	30	10
	contaminated.+PAH	0	0	10	0

concentrations adsorbed to the root ranged from 0.002 to 0.05 mg/g. A total of 15 equilibrations (three replicates of five initial concentrations) were made for each plant species.

Adsorption of naphthalene onto the roots was adequately described by the Freundlich isotherm:

$$q = K_f \, c^{1/n}$$

Freundlich parameters (K_f and $1/n$) and correlation coefficients (r^2) for fescue and alfalfa are given in Table III. Alfalfa had a greater affinity for naphthalene than fescue as indicated by their respective K_f values. The $1/n$ values were statistically equivalent. The highly significant r^2 values indicated that the Freundlich isotherm adequately described the adsorption behavior. However, the range of solution concentrations was fairly narrow and did not rigorously test the applicability of the isotherm.

If naphthalene and pyrene can be assumed to have similar adsorption characteristics, then it is possible to predict the total pyrene adsorbed in the greenhouse experiment. With root biomass in the greenhouse pots ranging from 1 to 3 g, the total mass adsorbed to the root surface would be nearly equal to the total pyrene uptake (Table II). Therefore, root uptake reported in Table II may be due totally to surface adsorption with very little assimilation into the interior of the roots.

Table III. Parameters for the Freundlich adsorption isotherms for adsorption of naphthalene on alfalfa and fescue roots

plant species	K_f	$1/n$	r^2
alfalfa	0.059	1.11	0.83
fescue	0.028	1.13	0.90
l.s.d. (P<0.05)	0.017	0.08	

Microbial Studies. Microbial counts for soils with alfalfa and no plants in the three soil treatments were obtained at planting (0 weeks), 4, and 24 weeks after planting. As shown in Table IV, microbial numbers significantly increased with time for all soil treatments planted with alfalfa. Without plants, microbial numbers increased slightly for uncontaminated soil and uncontaminated + PAH soil, and decreased for contaminated + PAH soil. Microbial counts in soils with plants were substantially higher than microbial numbers in soils without plants. The results do not show any significant difference between microbial numbers for each soil type even though microorganisms in the contaminated + PAH soil were acclimated to the target compounds.

Table IV. **Microbial colony forming units (cfu) in soils with and without plants**

soil treatment	treatment	0 weeks	4 weeks	24 weeks
		------------- cfu x 10^5 --------------		
uncontaminated	no plant	14	19	23
uncontaminated	plant	14	35	1640
l.s.d (P<0.05)		2	7	309
uncontaminated + PAH	no plant	13	17	17
uncontaminated + PAH	plant	13	33	1620
l.s.d. (P<0.05)		2	9	322
contaminated + PAH	no plant	7	5	6
contaminated + PAH	plant	7	16	1400
l.s.d. (P<0.05)		2	3	312

Dissipation of Pyrene in Contaminated Soil. Changes in pyrene concentrations demonstrated the typical pattern for soil bioremediation: very rapid dissipation in the early stages followed by slower rates. After 4 weeks, concentrations had declined from an initial level of 100 mg/kg to ≤2.2 mg/kg in the uncontaminated + PAH and ≤ 12.6 mg/kg in the contaminated + PAH soils (Table V). By 24 weeks, concentrations were ≤0.22 mg/kg in the uncontaminated + PAH soils and ≤2.4 in the contaminated + PAH soils.

Pyrene degradation in the uncontaminated + PAH soils was significantly greater in the presence of plants than without plants (Table V). All species were equally effective at enhancing biodegradation. Only alfalfa in the contaminated + PAH soil did not have statistically smaller concentrations of pyrene in the soil after 24 weeks when compared to soils without plants using a 5% confidence level (P<0.05). However, the difference was significant using a 6% level (P<0.06).

Overall, degradation of pyrene was much faster in the uncontaminated+ PAH soil than in the contaminated+PAH soil. For example, at 4 weeks, pyrene concentrations ranged from 0.7 to 2.2 for the uncontaminated+PAH soil and from 9.9 to 12.6 mg/kg for the contaminated+PAH soil. By 24 weeks, the range was 0.1 to 0.3 for the uncontaminated+ PAH soil and 1 to 2.5 mg/kg for the contaminated+PAH soil. PAH concentrations were approximately 10 times higher in the contaminated+PAH soil, but the trends in degradation were similar to those in the uncontaminated+ PAH soil.

Throughout the greenhouse experiment, leachates were collected from selected pots and analyzed for PAHs. No PAHs were detectable (<10 g/L) in any of the leachates. Therefore, transport out of the soil by leaching was not a possible mechanism for soil dissipation.

A soil sample containing 100 mg/kg pyrene was sealed in a plastic bag and stored at 4 °C. The sample was kept at field moisture and analyzed periodically to determine the extent of irreversible adsorption and abiotic degradation. After 24 weeks, the concentration of pyrene in this sample was 102 ± 5 mg/kg. Therefore, neither non-extractable (bound) pyrene nor non-biological degradation could account for the observed changes in pyrene concentration in the greenhouse experiment.

Table V. Dissipation of pyrene in soils with or without plants

soil treatment	plant species	pyrene concentration	
		4 weeks	24 weeks
		--------------- mg/kg ---------------	
uncontaminated+PAH	none	2.18	0.22
	alfalfa	0.67	0.15
	fescue	0.75	0.15
	switch grass	1.15	0.16
	sudan grass	1.48	0.15
	l.s.d ($P<0.05$)	0.91	0.05
contaminated+PAH	none	12.6	2.36
	alfalfa	9.89	1.67
	fescue	10.2	1.49
	switch grass	10.8	1.48
	sudan grass	10.7	1.32
	l.s.d ($P<0.05$)	ns	0.79

Mineralization of [^{14}C]Pyrene. To simulate the effect of root exudates, soils amended with [^{14}C]pyrene (uniformly labeled) were incubated in the presence of organic acids normally found in the rhizosphere (formic, acetic, and succinic).

Soils were collected from the rhizosphere of alfalfa or from non-vegetated soils. Sterilized soil was used as a control. The production of $^{14}CO_2$ was highest in the rhizosphere soil periodically amended with organic acids (Table VI). The increased mineralization of pyrene was most likely the result of elevated microbial populations in response to the organic acid amendment.

The rhizosphere and non-rhizosphere soils maintained at optimum moisture content with distilled deionized water exhibited no significant differences in the production of $^{14}CO_2$. This suggests that the enhanced degradation of hazardous organics in the rhizosphere is dependent upon the continuous exudation of organic acids or other carbon sources. The sterilized soil maintained at optimum water content with distilled deionized water produced $^{14}CO_2$ at barely detectable concentrations.

Pyrene mineralization in all soil treatments was fairly low. The highest amount of $^{14}CO_2$ in Table VI accounted for only 0.17% of the total [^{14}C]pyrene added.

Table VI. Mineralization of ^{14}C-pyrene in sterilized soil (control), unamended non-rhizosphere soil, unamended rhizosphere soil, and rhizosphere soil amended with organic acids

treatment	5 days	20 days	60 days
	---------------- $^{14}CO_2$ (dpm) ---------------		
control (sterile)	336	1994	3617
non-rhizosphere soil	5082	9580	12740
rhizosphere soil	5142	9870	13600
rhiz. soil + organic acids	5803	12580	18450
l.s.d. (P<0.05)	1340	1630	2360

Summary and Conclusions

The fundamental premise behind vegetative bioremediation is that the root environment will stimulate microbiological activity and result in increased degradation. As indicated by plant biomass, the plants growing in the landfarmed, contaminated soil were healthy and supported a normal, active root system. Microbial counts in the soil were 70 to 200 times greater in the presence of plants than in the absence of plants. The actively growing and exuding roots provided an excellent environment for enhanced biodegradation of the organic contaminants.

Concentrations of pyrene in the soil declined rapidly in the first 4 weeks of the experiment from 100 mg/kg to as low as 0.7 mg/kg. By 24 weeks, concentrations ranged from 0.15 to 2.4 mg/kg. Concentrations after 24 weeks were significantly lower in all vegetated soils than unvegetated soil (P<0.06 for alfalfa, $P < 0.05$ for other plant species). Plant uptake of pyrene and adsorption onto the surface of the roots alone cannot explain the observed dissipation. Likewise, irreversible adsorption, abiotic degradation, and leaching of pyrene were not significant. Therefore, biological degradation by microbes in the rhizosphere is the most likely explanation for the enhanced dissipation.

To further test the hypothesis that root exudates can stimulate degradation of PAHs, [^{14}C]pyrene was added to soil and incubated in the presence and absence of organic acids. Complete mineralization of pyrene to $^{14}CO_2$ was small ($\leq 0.17\%$), but mineralization was 36% greater when organic acids were added than when organics were not added. The organic acids added (formic, acetic, and succinic) are typically found in the rhizosphere, and these results strongly support the notion of enhanced biodegradation in the rhizosphere resulting from root exudation.

Future Research. Many questions remain unanswered concerning the role of plants in remediation. The experiments described above were terminated after 24 weeks; much more information is needed about long-term effects of vegetation. This approach also needs to be evaluated under field conditions to ensure that potential biases introduced in the greenhouse and laboratory are removed. The zone of influence of various plant species needs to be evaluated further; the depth of soil to which plants enhance biodegradation of hazardous organics is not known. The bulk of the root mass is usually found in the upper 20 cm of the soil, but roots can extend as deep as several meters.

The use of vegetation in bioremediation of contaminated sites is attractive because it is inexpensive and passive. With few inputs and little management, a successful vegetation remediation system could be superior to many alternative clean-up techniques. In this study, the use of plants to enhance the dissipation of PAHs in soil contaminated with petroleum hydrocarbons was investigated, and the results were promising.

Literature Cited

1. Walton, B.T.; Anderson, T.A. *Cur. Opin. Biotech.*, **1992**, *3*, 267-270.
2. Bell, R.M. *Higher Plant Accumulation of Organic Pollutants from Soils*; EPA/600/SR-92/138;U.S. Environmental Protection Agency: Cincinnati, OH, **1992**, pp. 1-4.
3. Finlayson, D.G.; MacCarthy, H.R. In *Environmental Pollution by Pesticides*; Edwards, C.A., Ed.; Plenum Press: New York, **1973**, pp. 28-53.
4. Crowdy, S.H.; Jones, D.R. *Nature*, **1956**, *178*, 1165-1167.
5. Mottley, J; Kirkwood, R.C. *Weed Res.*, **1978**, *18*, 187-198.
6. Hancock, J.L.; Applegate, H.G.; Dodd, J.D. *Atmos. Environ.*, **1970**, *4*, 363-370.
7. Shabad, L.M.; Cohan, Yu.L. *Ach. Geschwultforsch*, **1972**, *40*, 237-246.
8. Shiraishi, Y. *J. Food Hyg. Soc., Japan*, **1975**, *16*, 178-192.
9. Siegfried, R. *Naturewissenschaften*, **1975**, *62*, 300-311.

10. Edwards, N.T. In *Polynuclear Aromatic Hydrocarbons: A Decade of Progress*; Cooke, M; Dennis, A.J., Eds.; Tenth International Symposium; Battelle Press: Columbus, OH, **1988**, pp 211-229.
11. Sims, R.C.; Overcash, M.R. *Residue Rev.*, **1983**, *88*, 1-68.
12. Edwards, N.T. *J. Environ. Qual.*, **1983**, *12*, 427-441.
13. Ryan, J.A.; Bell, R.M; Davidson, J.M.; O'Connor, G.A. *Chemosphere*, **1988**, *17*, 2299-2323.
14. Jones, K.C.; Grimmer, G.; Jacob, J.; Johnston, A.E. *Sci. Total Environ.*, **1989**, *78*, 117-130.
15. Paul, E.A.; Clark, F.E. *Soil Microbiology and Biochemistry*; Academic Press: San Diego, CA, **1989**; pp. 81-84.
16. Alexander, M. *Introduction to Soil Microbiology, 2nd Edition*; John Wiley and Sons: New York, **1977**; pp. 423-437.
17. Tames, R.S.; Hance, R.J. *Plant and Soil*, **1969**, 30, 221-226.
18. Lappin, H.M.; Greaves, M.P.; Slater, J.H. *Appl. Environ. Microbiol.*, **1985**, *49*, 429-433.
19. Sandmann, E.; Loos, M.A. *Chemosphere*, **1984**, *13*, 1073-1084.
20. Walton, B.T.; Anderson, T.A. *Appl. Environ. Microbiol.*, **1990**, 56, 1012-1016.
21. Aprill, W., Sims, R.C. *Chemosphere*, **1990**, *20*, 253-265.
22. Al-Assi, A.A. *Uptake of Polynuclear Aromatic Hydrocarbons by Alfalfa and Fescue*, M.S. Thesis, Department of Civil Engineering, Kansas State University, Manhattan; **1993**.
23. Reilly, K. *Dissipation of Anthracene and Pyrene in the Rhizosphere*, M.S. Thesis, Department of Civil Engineering, Kansas State University, Manhattan; **1993**.
24. Lee, E.; Banks. M.K. *J. Environ. Sci. Health* **1994**, (in press)

RECEIVED February 15, 1994

Chapter 13

Grass-Enhanced Bioremediation for Clay Soils Contaminated with Polynuclear Aromatic Hydrocarbons

X. Qiu[1], S. I. Shah[1], E. W. Kendall[2], D. L. Sorensen[3], R. C. Sims[3], and M. C. Engelke[4]

[1]Central Research and Engineering Technology, Union Carbide Corporation, South Charleston, WV 25303
[2]Department of Environmental Protection, Seadrift Plant, Union Carbide Corporation, North Seadrift, TX 77983
[3]Department of Civil and Environmental Engineering, Utah State University, Logan, UT 84322
[4]Texas Agricultural Experiment Station, Research and Extension Center at Dallas, Texas A&M University, Dallas, TX 75252

In situ biotreatment of polynuclear aromatic hydrocarbon (PAH) contaminated clay soils is a challenge. A grass-enhanced bioremediation system was conceived as a relatively passive but potentially effective remediation method. Our laboratory studies to date have shown that a statistically significant increase in PAH removal is possible through the use of a grass-enhanced system. A field demonstration for grass-enhanced bioremediation for PAH contaminated soil is being conducted for a 2-yr period at a Union Carbide Plant. The objective is to evaluate, at pilot-scale level, the ability of a fibrous root system to facilitate the bioremediation of PAH in clay soils typical in the Gulf Coast area. The performance of the process is being evaluated by comparison of the PAH concentration reductions in soil planted with 'Prairie' Buffalograss to that in soil in the absence of grass. This pilot study will determine degradation kinetics of PAH compounds, formation of toxic metabolites, and the influence of soil characteristics and microbial populations in the field. Soil gas and solid phase of the top soil zone as well as the groundwater in the upmost aquifer are being monitored. A parallel experiment is being conducted to assess the performance of twelve warm-season grass species with various genetic origins. Randomized, complete block experiments were designed to appraise the effect of individual grass species. The visual differences in grass germination, establishment, survival, density, height, and root depth are being observed.

Polynuclear aromatic hydrocarbons (PAHs) in low permeability clay soils are difficult to treat by conventional in situ biotechnology. PAHs partition into clay-soil organic matter complexes and oil globules trapped in soil pores, becoming unavailable for biodegra-

0097–6156/94/0563–0142$08.00/0

dation. The low flux of nutrients and electron acceptors through the low permeable clay soil may also restrict microbial activities. PAHs are generally resistant to biodegradation at low redox potential (under methanogenic or sulfate reducing conditions) typical of contaminated subsurface environments (*1*). Maintaining well aerated conditions in low permeable soil is a challenge. Furthermore, bacteria and fungi cannot use high molecular weight PAHs as a sole carbon source (*2,3*). A readily degradable organic carbon source is required for cometabolism of high molecular weight PAHs.

PAHs are frequently observed in Union Carbide contaminated sites. Most PAH-contaminated soils contain a high percentage of clay, often greater than 50%. Union Carbide intends to develop a grass-enhanced soil bioremediation technology which would be low-cost, low-maintenance, effective at specific sites, and would avoid air pollution and LDR (land disposal restriction) impacts. Union Carbide and Utah State University jointly initiated a research in 1989. Since then, a series of laboratory studies has been conducted. Under laboratory conditions, the rates of PAH degradation in soil were enhanced in the presence of prairie grasses. Based upon the results, a field pilot-scale study was initiated. The primary purpose of the pilot study is to evaluate the enhancement of the rate and extent of PAH biodegradation in the presence of a grass root zone in the field. The second purpose is to identify competitive grass species with high potential for degrading PAHs in situ. The field scale evaluation is undergoing its first year of operation. Preliminary results of the field study are reported here.

Background

Deep Fibrous Root Zone Effects. The deep fibrous root system of prairie grasses may improve aeration in soil and provide a means for bringing microbes into contact with PAH compounds through the large surface of the root hair-soil interface. The zone of increased microbial activity and biomass at root-soil interface is called the rhizosphere. The carbohydrates, amino acids, etc. exuded from roots sustain a dense microbial community in the rhizosphere, which may enhance biodegradation rates of organic contaminants (*4,5*). The consortium of bacteria and fungi associated with the rhizosphere may possess highly versatile metabolic capabilities. In addition, organic exudates from roots may induce microbial cometabolism of high molecular weight PAHs.

Plant Uptake of PAHs. Many plant roots and shoots are capable of concentrating and detoxifying soil-borne chemicals such as pesticides and petroleum compounds.

Concentration Factor. The concentration factor is defined as the ratio of the chemical concentration in roots (or shoots) to that in soil water. High plant concentration factors for PAHs have been reported (*6*). Plant roots accumulate soil-borne pollutants with high octanol-water partition coefficient (K_{OW}) values to concentrations many times those found in the soil (*7*). PAHs are most likely to be accumulated in the roots and not be translocated to plant shoots because of their high log K_{OW} values (4.07 - 7.66) (*6,8*). Plant accumulation of PAHs provides an alternative pathway to remediate soils; however, the rate of such treatment may be slow. In PAH-contaminated soils, a PAH is partitioned between clay/soil organic matter and the liquid phase. The plant root can only sorb the compound in the liquid phase. PAHs have a high tendency to be sorbed on soil organic matter. The process of PAH dissolution and the subsequent partitioning into plant roots to reach equilibrium is slow.

Degradation and Polymerization of PAHs in Plants. Plants can use a variety of reactions to degrade complex aromatic structures to more simple derivatives. Aromatic ring-cleavage reactions in plant tissues, which lead to complete catabolism of the aromatic nuclei to carbon dioxide, have been reported (*9*). Benzo[a]pyrene (B[a]P), a five-ring PAH, can be metabolized into oxygenated derivatives in plant tissues (10). Although some of these derivatives are known to be more toxic than the original compounds, they appear to be polymerized into the insoluble plant lignin fraction, which may be an important mechanism for their detoxification. With plant seedlings, B[a]P was assimilated into organic acids and amino acids (11). Complete degradation of B[a]P to carbon dioxide was also observed with a wide range of plants.

Short-Term Effects. Cleanup of PAH contaminated soils seems possible with the aid of grasses. While PAH degradation and polymerization may take considerable time, turf grasses provide short-term benefits by reducing the risk posed to human health and the environment. Dense turf grasses serve as a botanical cap to prevent exposure to soil-borne contaminants by direct ingestion, inhalation, and skin contact. Turf grasses also reduce potential contaminant migration through surface runoff and infiltration to groundwater.

Methodology

PAH degradation in sandy loam soils was enhanced by grasses under laboratory conditions. On the basis of laboratory results, the field pilot study was initiated. The approach of the field scale study includes: (1) selecting a representative demonstration site in a chemical manufacturing area, (2) evaluating the treatment performance by assessing degradation kinetics of PAH compounds at field scale, (3) applying a quality assurance/quality control plan to site characterization and performance evaluation tests, (4) applying statistics to the sampling and analysis plan and for evaluation of field scale variability in treatment, and (5) identifying competitive grass species by evaluating the performance of a variety of grasses.

Laboratory Studies. The preliminary laboratory study showed that the concentration reduction of high molecular weight PAHs in a sandy loam soil was significantly greater (minimum of 90% confidence) in the presence of prairie grasses and cow manure when compared to unvegetated controls (12). The eight prairie grasses tested were Big bluestem *Andropogon gerardi*, Indiangrass *Sorghastrum nutans*, Switchgrass *Panicum virgatum*, Canada wild-rye *Elymus canadensis*, Little bluestem *Schizachyrium scoparius*, Side oats grama *Bouteloua curtipendula*, Western wheatgrass *Agropyron smithii*, and Blue grama *Bouteloua gracilis*. With an initial spiked concentration of 10 mg-PAH/kg-soil, vegetation mediated a 97.3% reduction in benz[a]anthracene, a 93.6% reduction in chrysene, a 88.3% reduction in benzo[a]pyrene, and a 45.3% reduction in dibenz[a,h]anthracene after 219 days of incubation. The study also indicated a significant reduction of water infiltration through the grass root zone.

Subsequent laboratory mass-balance experiments, using radiolabelled compounds, indicated that mineralization rates of phenanthrene and pentachlorophenol (PCP) in a sandy loam soil were significantly enhanced in the presence of Crested wheatgrass *Agropyro cristatum* (13,14). With an initial spiked concentration of 100 mg/kg-soil, the phenanthrene mineralization by the plant system was 37% in two weeks compared to 7% by the control system. The fate of PCP spiked onto a sandy loam soil at a concentration of 100 mg/kg was evaluated in the presence of Crested wheatgrass. In 155 days the mineralization of PCP was 22% compared to 6% in a control system. For the planted system, 36% of PCP was associated with plant tissue (21% with roots and 15% with shoots), and 33% was remained in the soil and leachate. Volatilization rate of naphthalene in a sandy loam soil was enhanced in the presence of Bell's Rhodesgrass *chloris gayanaprairie* , which suggested that the grass roots improved soil aeration (15).

A preliminary test indicated that no significant detrimental effects were caused by the presence of PAHs on seed germination and root radical elongation for five prairie grasses (*12*). A subsequent Buffalograss seed germination test indicated that the grass was tolerant to high PAH concentrations (>300 mg/kg-soil) and may survive in clay soil with significant levels of PAHs, oil and grease (Sorensen, D. L., personal communication, Utah Water Res. Lab., Utah State Univ., 1992).

Site Selection and Characterization. Site selection and characterization were based on available information regarding regional hydrogeology, contamination history, background soil sampling and analysis, as well as long-term land use planning and safety considerations.

Site Selection. Soil samples were taken at 0.3 to 0.9-meter depths below the ground surface from ten candidate sites. Composite soil samples for each of the ten candidate sites were analyzed for PAH constituents. The final site was selected based on its PAH constituents and concentration levels.

Contamination History. The pilot study plot was constructed between the subgrade foundations of an old olefins unit building at Union Carbide's Seadrift Plant. The building had been demolished after operation was shut down in 1983. The soil underneath the building had been contaminated by process material, mainly pyrolysis gasoline, since 1954. The contaminants present in the soil were basically aromatic compounds and PAHs.

Geological and Climatic Description. The Seadrift Plant is located in Calhoun County, a coastal area of Texas. The test site is a topographically leveled surface at an elevation of approximately 9 meters above MSL (Mediterranean Sea Level). The groundwater table occurs approximately 2 meters below the surface. The topsoil is fine clay called "Texas Gumbo". The top of the uppermost sandy aquifer is approximately 4 meters below the ground surface. The groundwater is not suitable for drinking because of its high salinity.

Local weather records (1951-1980) showed that the mean annual precipitation was 100.3 cm and the mean annual lake evaporation was 149.9 cm. The annual average daily low and high temperatures were 17°C and 26°C, respectively. The monthly extremes were 7°C in January and 33°C in July and August.

Site Characterization. Background sampling and analysis of soil samples were performed to determine soil characteristics and contaminant composition and distribution. Details are presented in subsequent sections of this paper.

Experimental Design of Remedial Performance Evaluation. Three test plots were constructed for the evaluation. Plot 1 was an unvegetated control. Plot 2 was planted with 'Prairie' Buffalograss (*Buchloe dactyloides)*. Plot 3 was constructed to screen the performance of different grass species. Performance evaluation of grass enhanced bioremediation involves the comparison of extractable PAH concentration reduction in soil with time and depth between the testing Plot 1 and Plot 2. The experimental design for Plot 3 is presented in a subsequent section of this paper.

Factorial Design of Experiments. Table I shows a three-factor, factorial experimental design for Plots 1 and 2 to evaluate the performance of grass-enhanced PAH biodegradation.

TABLE I.   Experimental design for Plot 1 and Plot 2

Source	Description	Number
Replicates		$\leq 10^a$
Factor A: Vegetation condition		2
	Plot 1 - Unvegetated control	
	Plot 2 - Vegetated	
Factor B: Sample depth		2
	Surface (0 - 0.3 m)	
	Subsurface (0.3 - 1.2 m)	
Factor C: Time	Once per 4 - 6 months[a]	7
Total soil samples		≤ 280

[a] See sampling strategy in the Section "Sampling and analysis plan".

Sampling and Analysis Plan. A sampling and analysis plan (SAP) was developed for site characterization and remedial performance evaluation. The primary purpose of the SAP was to determine the PAH concentration reduction.

Variability. Field scale data variability was evaluated for an accurate estimate of PAH biodegradation rates. Physical and chemical properties of soils are rarely homogenous throughout a site. The variability of these properties may range from 1 to more than 100 percent of the mean value within relatively small areas[16]. Contaminant concentrations often have the highest variability. The total variance in field data can be approximately defined by Equation 1.

$$V_t = V_s / n + V_a / k*n \qquad [1]$$

where, n = the number of samples, k = the number of analyses per sample, Vt = the total variance, Va = the analytical variance, and Vs = the sample variance. Analytical procedures frequently achieve a precision level of 1-10%, while soil sampling variation may be greater than 35% (16). Sampling should be designed to reduce the magnitude of Vs wherever possible. In general, sampling efforts to minimize Vt will result in the best precision. A systematic randomized sampling strategy was employed in this field study to reduce the sample variance.

Systematic Randomized Sampling Strategy. The number of samples to be taken from the plots should be sufficient to accurately estimate the variation of contaminant concentration. In addition, sampling should be designed to minimize altering soil environmental factors which interfere with plant growth, soil moisture dynamics, and gas exchange. A limit of disturbance of less than 1% of the plot surface, by sampling, was chosen to protect against these interferences. The number of sampling points chosen on each plot was no more than ten per each sampling event.

The number of sampling points was decided based on the expected mean and variance of the contaminant concentration and the level of error that can be tolerated in the test data. Equation 2 was used to calculate the appropriate number of samples (17).

$$n = t^2 (CV)^2 / p^2 \qquad [2]$$

where, n = appropriate number of samples, t = the Student's t for the specified confidence, CV = s/x, the coefficient of variation, s = standard deviation, x = expected mean, and p = the allowed margin of error (%).

Figure 1 illustrates the grid layout for a test plot. Each cell contains a single sample point. The coordinates (a, b) for the sample points within the cell are chosen at random. The aligned square grid (dashed lines) design preserves a pattern of regular spacing between sample points. Table II shows the (a, b) coordinates for each sampling event. The total number of sampling points for the subsequent sampling events may change depending on the spatial variation found in the previous sampling event. The concept of designating the sampling locations will be the same.

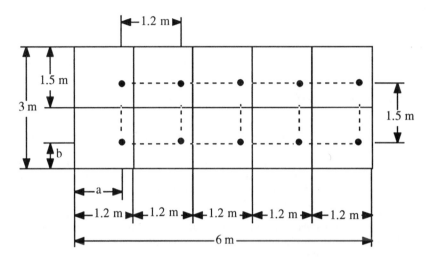

FIGURE 1.   Sampling Points Layout for Plots 1 and 2

TABLE II.   The Randomized Coordinates For Sampling Locations

Sampling Event (Event #)	Coordinate "a" (Inches)	(cm)	Coordinate "b" (Inches)	(cm)
I	28	71	17	43
II	06	15	18	46
III	07	18	52	132
IV	24	61	30	76
V	35	89	22	56
VI	45	114	11	28
VII	19	48	05	13

Sampling Frequency. The establishment of the grass root system and the microflora within the rhizosphere may take considerable time. The initial difference of PAH concentration reduction between vegetated and non-vegetated plots may not be significant. Therefore, sufficient time was allowed before intensive sampling. A sufficient number of samples per each event is required to allow statistical analysis of spatial variations. Fewer sampling events and more samples per event were planned for cost effectiveness. Sampling was tentatively scheduled for once every 4 to 6 months during the first two years. The schedule may need to be adjusted in response to the kinetics of degradation observed and the rate and depth of root penetration.

Parameters measured. Table III contains the parameters measured in the field study. The five categories of parameters are contaminant constituents, soil physicochemical characteristics, soil aeration condition, microbial population, and grass growth. Some of the parameters are to be analyzed in initial and final samples only. The details of the analytical methods listed on the table can be found in "EPA SW846 Test Method for Evaluating Solid Waste" and "Methods for Soil Analysis" (*18, 19*).

Groundwater and Soil Gas Monitoring. The uppermost aquifer groundwater in up- and down-gradient wells is being monitored to evaluate potential contaminant migration. Five soil gas probes at 0.6 or 1.2 m deep were installed in test plots. Soil gas samples are taken by a vacuum pump and analyzed for CH_4, CO_2, and O_2 by GC/TCD (gas chromatography with thermal conductivity detector) and oxygen analyzer. Soil gas composition indicates the soil aeration condition.

Sample Collection, Packing, and Shipping. A power-driven Shelby tube sampler was used for soil sampling. For each sampling point, surface (0 - 0.3 m depth) and subsurface (0.3 - 1.2 m depth) soil core samples were collected. The soil core was bisected radially on a clean plastic sheet. The half soil cores were then cut into approximately 3 cm long pieces. A fraction of the soil from each piece was immediately transferred into a clean glass container with a Teflon lined cap. Each sample jar contained approximately 500 grams of soil. Shelby tubes, cutting knives, and latex gloves were detergent and water washed between samples to minimize cross contamination. Groundwater samples were taken from the monitoring wells after half an hour purging. Sample containers were packed in a cooler and shipped to the Utah Water Research Laboratory, Utah State University by overnight transportation.

Samples were stored at 5°C until analysis, but no more than two weeks. Back-up soil and groundwater samples were sent to Union Carbide Tech Center Research lab and stored at 5°C.

Identification of Competitive Grass Species. Competitive grass species must have a deep and fibrous root system, be suitable for tight clay soils, and require minimal maintenance. In addition, competitive grasses must have high potential for PAH sorption and degradation, and be able to tolerate high concentration of PAH and other contaminants. The competitive grass species are to be selected from a broad range of species as part of the field performance evaluation.

'Prairie' Buffalograss. 'Prairie' Buffalograss was planted in Plot 2. 'Prairie' Buffalograss is cold-, heat-, and drought-resistant and highly competitive in marginal soils and requires minimal maintenance. Buffalograss evolved throughout the central and southern Great Plains of the United States and therefore has excellent biological traits for adaptation to the test site. 'Prairie' is vegetatively propagated with fine dense texture and an extensive, fibrous, deep, root system. Under normal management practice the grass will attain a mature non-mowed plant height from 10 to 15 cm 'Prairie' Buffalograss will perform best on soil with a high clay content and a neutral to alkaline soil pH. Once established, 'Prairie' Buffalograss requires little or no supplemental irrigation.

TABLE III. Parameters Measured

Parameter	Matrix	Event	Methods
Contaminant constituents:			
Naphthalene	Soil, groundwater & grass tissue	all	Soxhlet extraction; HPLC or GC/MS[b]
2-methylnaphthalene	-"-	-"-	-"-
biphenyl	-"-	-"-	-"-
Acenaphthylene	-"-	-"-	-"-
Fluorene	-"-	-"-	-"-
Phenanthrene	-"-	-"-	-"-
Anthracene	-"-	-"-	-"-
Fluoranthene	-"-	-"-	-"-
Benzidine	-"-	-"-	-"-
Pyrene	-"-	-"-	-"-
Benzo[a]anthracene	-"-	-"-	-"-
Chrysene	-"-	-"-	-"-
Benzo[b]fluoranthene	-"-	-"-	-"-
Benzo[k]fluoranthene	-"-	-"-	-"-
Benzo[a]pyrene	-"-	-"-	-"-
Indeno(1,2,3-cd)pyrene	-"-	-"-	-"-
Benzo[g,h,i]perylene	-"-	-"-	-"-
Indicative reaction intermediates	-"-	-"-	Soxhlet extraction; GC/MS
Soil aeration condition:			
O_2	Soil gas		Oxygen analyzer
CO_2 & CH_4	-"-		GC/TCD[c]
Soil characteristics:			
Texture	Soil	Initial	Hydrometer
Cation exchange capacity	-"-	-"-	
Cl^-, SO_4^{2-}	-"-	-"-	IC[d]
Na^+, Ca^{2+}, Mg^{2+}	-"-	-"-	ICP[e]
TKN, NH_4^+, NO_3^-	-"-	all	Dist.; Color; IC
Total P, soluble PO_4^{3-}	-"-	-"-	Color; IC
pH	-"-	-"-	Electrode
Total organic carbon (TOC)	-"-	-"-	TOC analyzer
Microbial Populations:			
Viable aerobic heterotrophic bacteria	Soil & roots	Initial/ final	Plate Count
Viable fungal propagules	-"-	-"-	-"-
Viable naphthalene utilizing bacteria	-"-	all	Plate count/Activity assay
Grass growth:			
Overall appearance	Grass	all	Visual observation
Density, height	-"-	-"-	-"-
Root density and penetration depth	-"-	-"-	Estimate from soil core

[b] High performance liquid chromatography or gas chromatographic/mass spectroscopy

[c] Gas chromatography with thermal conductivity detector

[d] Ion chromatography

[e] Inductively coupled argon plasma spectroscopy

Twelve Grass Species being Evaluated in Plot 3. The effects of a grass root zone system on PAH removal may be species-dependent. Each specific grass species influences characteristics of the grass root zone, including: (1) the microbial diversity and metabolic capability associated with the rhizosphere, (2) the highly specialized, soil-inhabiting fungi in mycorrhizal association, (3) the characteristics of organic exudates from grass roots, and (4) the texture and depth of the root zone. Twelve warm season grass species from different genetic origins are being tested on Plot 3. These are naturally adapted to the Great Plains and Coastal areas of Texas. Table IV lists their names and application forms.

TABLE IV.   Twelve Species To Be Tested On Plot 3

ID	Name	Application Form
1	Seashore Paspalum *Paspalum vaginatum* var. 'Adalayd'	Sod
2	Buffalograss *Buchloe dactyloides* var. 'Prairie'	Sod
3	Bermudagrass *Cynodon dactylon* var. 'Texturf 10'	Sod
4	Zoysiagrass *Zoysia japonica* var. 'Meyer'	Sod
5	St. Augustinegrass *Stenotaphrum secundatum* Raleigh	Sod
6	K.R. (King Ranch) Bluestem *Bothriochioa ischaemum*	Seed
7	Common Buffalograss *Buchloe dactyloides*	Seed
8	Weeping Lovegrass *Eragrostis curvula*	Seed
9	Kleingrass *Panicum coloratum* var. 'Verde'	Seed
10	Blue grama grass *Bouteloua gracilis*	Seed
11	Texas Bluebonnet *Lupinus texensis*	Seed
12	Winecup *Callirhoe involucrata*	Seed

Tall grasses. Short grasses are preferred to ensure safety in chemical manufacturing areas. Tall grasses may increase hazards of fire and hide snakes. Most grasses selected are short species except Weeping Lovegrass and Kleingrass (0.9 m - 1.2 m). Weeping Lovegrass is a vigorous bunch grass with a deep fibrous root system. It is well adapted to a wide range of soil types (acidic to alkaline) and highly competitive against weeds. 'Verde' Kleingrass is a bunch-type forage grass with good fibrous root system, generally well adapted to a broad range of soil conditions. It is a prolific seed producer developed specifically for its potential use as bird seed to attract game birds in pasture and reclamation sites.

Sodded and seeded Buffalograsses. Both sodded and seeded cultivars of Buffalograss are being tested to assess the importance of inoculum of microbes which are already symbiotic with the grass sods. Seeded Buffalograss does not have the competitive ability of sodded 'Prairie' Buffalograss; however, it possesses most other biological attributes of the genus.

Other sodded grasses. Bermudagrass is well adapted to low maintenance conditions. It is highly competitive and develops a deep fibrous root system. Zoysiagrass is adapted to broad conditions including extremes in soil pH (4.0 - 9.0), relatively low water and nutritional requirements. Once established and mature, the turf will be low-growing and highly competitive against weeds. St. Augustinegrass is an aggressive warm season turf with an extensive coarse leaf texture and a coarse deep rooting system. It is well adapted to a broad range of soil types, relatively low growing with a moderate to high water and nutritional requirement.

Other seeded grasses. K. R. Bluestem is well acclimated to the road sides and marginal distributed soils of South Texas. This species is well adapted to minimal maintenance and is an important species in the conservation and stabilization of Texas land resources. Bluegrama is truly a native short grass prairie plant. It is bunch type with a low-growing fibrous root system.

Natural flowers. Texas Bluebonnet and Winecup are natural flowers, which are aesthetically pleasing at remediation sites. The flowers were planted mixed with Buffalograss or Bluegrama. Using forbs affords us the opportunity to examine the potential for plants other than those in the grass family. Texas Bluebonnet is a legume (nitrogen fixing plant) and should stimulate microbial activity (*20*).

Experimental Design. The experiments on Plot 3 were designed to assess the performance of the twelve grass species. The assessment is based on visual observation. Frequent sampling and analysis are not required. After 2 years, a composite sample from each subplot will be analyzed to complete the assessment.

Plot 3 is divided into 36 subplots with a size of 1.5 m x 1.5 m. Figure 2 shows a randomized complete block design with three replicates. The number marked on each subplot refers to the ID number of the species listed in Table IV.

Visual observations were conducted plot by plot with 30 cm x 30 cm as a unit to evaluate: (1) rate of seed germination, (2) percentage survival of seedlings or solid sodding, (3) grass density and height, and (4) extent of root zone establishment. Appraisal is based on a nine grade system. A statistical analysis of the visual observations in conjunction with final soil core analysis will be performed by the end of the two-year test.

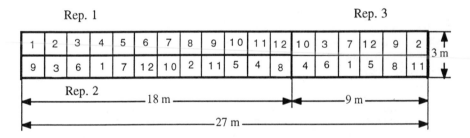

FIGURE 2. Randomized Complete Block Design Of Plot 3.
Numbers within blocks correspond to the ID numbers in Table IV.

Results

Soil Characteristics. Soil samples were analyzed for soil texture, physicochemical properties, and nutrient status. The results are presented in Table V. The soil is clay type with more than 50-60% clay content. The organic carbon content is approximately 2%. The cation exchange capacity (CEC) was found to be relatively high at approximately 40 meq/100 g, which indicates a high sorption capacity of the soil. The soil is neutral to weakly basic with a pH value of 7.5. The nutrient status of the soil seemed to be suitable for vegetation. The soil has relatively high calcium concentration (7800 mg/kg-soil), which causes phosphorus precipitation. The high sulfate concentration (205 mg/kg-soil) indicated that the black colored soil may be under a sulfate reducing condition. The soil turned gray after exposure to air.

TABLE V.   Physicochemical Characteristics of the Soil

Characteristic	CHAR[f]	Sample COMP[g]
Texture (Hydrometer)	Clay	Clay
% Sand	29	25
% Silt	20	15
% Clay	51	61
Organic Carbon (%)	2.6	1.3
Cation Exchange Capacity (meq/100 g)	42.7	34.0
pH	7.5	7.6
Cations (mg/kg)[c]		
Na^+	84	72
Ca^{2+}	7800	7500
Mg^{2+}	3	3
Anions (mg/kg)[c]		
Cl^-	150	NA[e]
SO_4^{2-}	205	120
HCO_3^-	16300	17600
Nutrients		
Total Kjeldahl N (%)	0.05	0.05
NO_3-N (mg/kg)[h]	90	13
NH_4^+ (mg/kg)[i]	< 5	< 5
Total phosphorus (%)	0.007	0.009
Soluble PO_4^{3-} (mg/kg)	< 1	< 1

[f] CHAR = Soil sample collected from Plot 2
[g] COMP = Composite sample made up of samples taken from Plot 1 and Plot 3
[h] 1:10, soil:water, extractable; [d] 1:10, soil:1N KCl, extractable
[i] NA = Not analyzed

Contaminant Analysis. Normal and lognormal probability plots for contaminant concentration distribution were developed for the background soil data set (21). The correlation coefficient was tested against the null hypothesis of normality or lognormality. The soil concentrations of PAHs were not normally distributed but lognormally distributed. The geometric mean with its corresponding confidence limit was the appropriate estimate for the site soil concentrations of PAHs. The mean and standard deviation for PAH concentrations in background soil samples are summarized in Table VI. Figure 3 shows the spatial concentration distribution of total PAHs in soil.

The average contaminant concentrations in surface soils were generally higher than those in subsurface samples. Analysis of variance showed that PAH compound concentrations in surface soil samples were significantly higher than those in subsurface soils, except acenaphthene. There were no significant difference between PAH concentrations of plots 1 and 2, except for naphthalene. Naphthalene concentration in Plot 1 was significantly higher than that in Plot 2 (21).

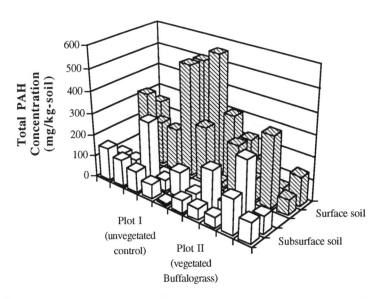

FIGURE 3. Spatial Variation Of Total PAH Concentrations in Soil
The columns on the left side represent the sampling points of unvegetated control plot 1. The columns on the right side represent the sampling points on Buffalograss vegetated plot 2. The shaded columns are concentrations in surface (0 - 0.3 m below ground surface) soil samples. The blank columns are concentrations in subsurface (0.3 - 1.2 m below ground surface) soil samples.

Concentration Variation and Appropriate Number of Samples. As identified in the sampling and analysis plan, the appropriate number of samples depends on the concentration variation. Table VI shows that the coefficient of variance for PAH concentrations ranged from 51% - 119% in surface soil samples and 42% - 128% in subsurface soil samples.

The appropriate numbers of samples (n) in plots 1 and 2 for surface and subsurface soil were calculated (Table VI) based on a ±40% of error allowance (p). The total number of sample points per plot, needed to meet the goal, ranged from 6 to 52 at 95% confidence level for the five target PAHs. However, because of the limited plot size, the total sample points per plot for each event should not exceed ten (<1% of the plot surface) so as not to interfere with the test. Based on ten soil core samples per plot (n' = 10), the error margin (p') that can be achieved varied from ±36% to ±85% for surface soil and from ±30% to ±91% for subsurface soil. The precision of concentration measurement for PAHs could hardly meet a ±40% error allowance. The error of concentration measurement was less than ±60% for acenaphthylene, acenaphthene, and anthracene, ±70% for phenanthrene, and 91±% for naphthalene.

TABLE VI.   Concentration Variation and Appropriate Number of Samples

Plot ID		Naphthalene (mg/kg-soil)		Acenaphthylene (mg/kg-soil)		Acenaphthene (mg/kg-soil)	
		Surface	Subsurface	Surface	Subsurface	Surface	Subsurface
1	Mean 'x'	206.14	56.25	8.40	3.86	5.97	3.80
	Count	10	10	10	10	10	10
	Std. dev. 's"	154.61	72.00	4.28	1.62	3.82	2.58
	CV	0.75	1.28	0.51	0.42	0.64	0.68
	p	0.40	0.40	0.40	0.40	0.40	0.40
	Student 't'	2.26	2.26	2.26	2.26	2.26	2.26
	$n = t^2 CV^2/p^2$	17.96	52.30	8.30	5.63	13.08	14.76
	n'	10	10	10	10	10	10
	p'	0.54	0.91	0.36	0.30	0.46	0.49
2	Mean 'x'	70.61	55.53	7.26	3.86	4.63	4.85
	Count	10	10	10	10	10	10
	Std. dev. 's"	84.03	45.53	4.36	1.62	2.45	3.69
	CV = s/x	1.19	0.82	0.60	0.42	0.53	0.76
	p	0.40	0.40	0.40	0.40	0.40	0.40
	Student 't'	2.26	2.26	2.26	2.26	2.26	2.26
	$n = t^2 CV^2/p^2$	45.21	21.46	11.49	5.63	8.97	18.44
	n'	10	10	10	10	10	10
	p'	0.85	0.59	0.43	0.30	0.38	0.54

TABLE VI.   Concentration Variation and Appropriate Number of Samples (cont')

Plot ID		Phenanthrene (mg/kg-soil)		Anthracene (mg/kg-soil)		Total PAHs (mg/kg-soil)	
		Surface	Subsurface	Surface	Subsurface	Surface	Subsurface
1	Mean 'x'	15.15	6.96	3.82	2.74	245.80	82.97
	Count	10	10	10	10	10	10
	Std. dev. 's"	12.88	4.80	3.21	1.37	169.60	73.84
	CV	0.85	0.69	0.84	0.50	0.69	0.89
	p	0.40	0.40	0.40	0.40	0.40	0.40
	Student 't'	2.26	2.26	2.26	2.26	2.26	2.26
	$n = t^2 CV^2/p^2$	23.06	15.20	22.52	7.98	15.20	25.29
	n'	10	10	10	10	10	10
	p'	0.61	0.49	0.60	0.36	0.49	0.64
2	Mean 'x'	17.43	12.05	4.93	2.82	125.94	87.75
	Count	10	10	10	10	10	20
	Std. dev. 's"	16.91	10.97	3.99	1.38	102.01	57.04
	CV	0.97	0.91	0.81	0.49	0.81	0.65
	p	0.40	0.40	0.40	0.40	0.40	0.40
	Student 't'	2.26	2.26	2.26	2.26	2.26	2.26
	$n = t^2 CV^2/p^2$	30.04	26.44	20.94	7.66	20.94	13.49
	n'	10	10	10	10	10	10
	p'	0.69	0.65	0.58	0.35	0.58	0.46

Microbial Enumeration. Table VII shows the data for viable bacteria and fungi enumerated from the soil sample. The native viable bacteria and fungi in soil are representative of agricultural soil. The large population of naphthalene utilizing bacteria indicated that native bacteria are capable of degrading PAHs, provided the available nutrients and electron acceptors are not limited.

TABLE VII.   Microbial Enumeration of Background Soil Samples

Plot ID	Total Viable Bacteria		Colony Forming Units per Gram Fungal Propagules		Naphthalene Utilizing Bacteria	
	Surface	Subsurface	Surface	Subsurface	Surface	Subsurface
Plot 1 Mean	1.33E+07	1.04E+07	2.48E+04	4.44E+04	4.75E+06	1.93E+06
Std. dev.	2.11E+07	1.03E+07	3.95E+04	8.71E+04	8.01E+06	2.34E+06
Plot 2 Mean	6.33E+07	3.46E+07	3.68E+05	1.04E+06	1.91E+07	1.45E+07
Std. dev.	2.52E+07	4.88E+07	7.99E+05	2.21E+06	1.23E+07	2.18E+07
Plot3 Mean	7.65E+06	3.10E+05	1.65E+05	2.00E+02	2.28E+06	3.00E+05
Std. dev.	4.74E+06		1.91E+05		2.44E+06	

Discussion

The most popular way to characterize a site with regard to the contaminant concentration is to estimate an arithmetic mean concentration and some degree of variation associated with the mean. However, examination of the soil analysis data set from the grass plots indicated that soil concentrations of PAHs were shown to be lognormally distributed. The geometric mean, with its corresponding variance, was the most appropriate estimate for on-site soil concentrations. Calculating the appropriate sample number based on the normal distribution may not be appropriate for a specific contaminated site.

Statistical design was critical to a field scale soil remediation study. Quantitative data variability should be evaluated for an accurate estimate of PAH concentration reduction. Determination of site spatial concentration variability is a prerequisite for identifying sample number requirements. The number of soil samples should be sufficient to accurately estimate the variation of contaminant concentration. However, the number of samples may have to be limited to minimize disturbing soil environment. A compromise was made to meet both requirements. The site contamination was highly heterogeneous. The high variance in soil may cause the estimate of PAH concentrations less accurate (±70 - 90% error). A performance evaluation may not be reliable unless a significant concentration reduction is achieved. The field scale study may need longer time to fulfill an appropriate performance evaluation.

While PAH degradation in sandy loam soils was enhanced by grasses in laboratory reactors, clay soils in field situations represent a more difficult challenge. The site soil had approximately 2% organic carbon content (OC%) and a cation exchange capacity (CEC) of approximately 40 meq/100 g. The relatively high OC% and CEC values indicated a high sorption capacity. The site soil has been contaminated for more than 20 years. PAHs were firmly adsorbed onto the soil particle surface. The bioavailability of PAHs may be extremely low in such an aged contaminated soil. For soils freshly spiked with PAHs in laboratory reactors the contaminants would be expected to be more easily removed or bioavailable. In addition, decontamination of a clay soil is more difficult than the sandy soil used in laboratory studies due to poorer aeration and higher sorption (higher surface area) characteristics of the clay soil.

A large population of naphthalene utilizing bacteria in the site soil indicated that the native bacteria are capable of degrading PAHs, provided the nutrients and electron acceptors are available. A PAH acclimated microbial community seemed to be present at the site. The native microbial activity may be enhanced in the presence of grass root zone system due to the improved aeration condition and the available nutrients (amino acids and carbohydrates) in rhizosphere. It would be interesting to find out which consortia of bacteria may be effective for PAH degradation in the rhizosphere. It is also interesting to find out how specific is the rhizosphere effect. Is it highly plant specific?

Conclusions

The site soil concentrations of PAHs were lognormally distributed. The spatial variation of PAH concentrations in soil is significant. The coefficient of variation of PAH concentrations ranged from 42% to 128% in the site soil. The number of soil samples should be sufficient to reliably estimate the variation of contaminant concentration. However, the number of samples may have to be limited to minimize disturbing soil environment. For each sampling event ten soil sample cores were taken from each plot. The error in the concentration measurement for PAHs was less than ±70%, except for naphthalene. The error of naphthalene measurements was less than ±91%.

The site soil was neutral to slightly basic with relatively high calcium and sulfate concentrations. The soil had approximately 2% organic carbon content (OC%) and a cation exchange capacity (CEC) of approximately 40 meq/100 g, which indicated a high sorption capacity. The bioavailability of PAHs may be low. A PAH acclimated microbial community seemed to be present at the site. The native bacteria are capable of degrading PAHs, provided the nutrients and electron acceptors are available.

The rate and extent of PAH degradation with the aid of grasses and competitive grass species can be determined, upon completion of this two-year field test. Further research needs are to identify the specific microbial consortium acclimated for PAH degradation in the presence of particular grasses and the plant specific rhizosphere effects.

Acknowledgements

This study is a part of Union Carbide remediation related research program. The authors thank Joe Stubbs, Joe Padilla, et al., Union Carbide Seadrift Plant, for carrying out the field pilot study. Special thanks must be given to Mr. Jack R. Battalora for allowing us the opportunity to conduct the field demonstration. The authors also thank Mr. Peter Hannak for the helpful discussion and Trena Summerfield for the help of preparing the manuscript.

At Utah State University, Venubabu Epuri and Jim Herrick performed field sample analysis. Wayne Aprill, Ari Ferro, and Jeff Watkins conducted laboratory studies. Thanks are due to Mr. John Matthews, USEPA Kerr Laboratory, and Dr. Daniel F. Pope, Dynamac Corporation, for partially funding the studies.

Literature Cited

1. McFarland, M. J.; Sims, R. C. Thermodynamic framework for evaluating PAH degradation in the subsurface. *Journal of Groundwater*, **1991,** *29(6)*:885-698.
2. Keck, J.; Sims, R. C.; Coover, M.; Park, K.; Symons, B. Evidence for cooxidation of poly nuclear aromatic hydrocarbons in soil. *Water Res*, **1989,** *23(12)*:1467-1476.
3. Bulman, T.; Lesage, S.; Fowlie P. J. A.; Weber, M. D. The persistence of poly nuclear aromatic hydrocarbons in soil. *PACE Report ,* **1985,** *No. 85-2.* Petroleum association for conservation of the Canadian Environment, Ottawa, Canada.

4. Anderson, T. A., Walton, B. T. *Comparative plant uptake and microbial degradation of trichloroethylene in the rhizospheres of five plant species- implications for bioremediation of contaminated surface soils. ORNL/TM-12017,* Oak Ridge National Laboratory, Oak Ridge, Tennessee, 1992, 206 pp.
5. Fitter, A. H., Hay, R. K. M. *Environmental Physiology, 2nd Ed.* Academic Press, London, UK, 1987, 423 pp.
6. Ellwardt, P. Variation in content of polycyclic aromatic hydrocarbons soil and plants by using municipal wastes composts in agriculture. *Proc. Series - Soil organic matter studies.* **1977,** 2:291-298.
7. Bell, R. M. *Higher Plant Accumulation of Organic Pollutants from Soils.* EPA/600/R-92/138; NTIS: Springfield, VA, 1992.
8. Harms, H.; Sauerbeck, D. Toxic organic compounds in municipal waste materials: Origin, content, and turnover in soils and plants, *Angew. Botanik.* **1984,** *58:* 97-108.
9. Ellis, B. E. Degradation of aromatic compounds in plants. *Lloydia.* **1974,** *37(2):*168-184,.
10. Harms, H.; Dehren, W.; Monch, W. Benzo[a]pyrene metabolites formed plant cells. *Z. Naturforsch.* **1977,** *32:* 321-326.
11. Sims, R. C.; Overcash, M. R. Fate of polynuclear aromatic hydrocarbons (PNA's) in soil-plant system. *Residue Revs.* **1983,** *88:* 2 - 68.
12. April, W, Sims, R. C. Evaluation of the use of prairie grasses for stimulating polyuclic aromatic hydrocarbon treatment in soil. *Chemosphere* .**1990,** *20(1-2):*253-265.
13. Ferro, A. M. *Biodegradation of Phenanthrene and pentachlorophenol mediated by vegetation.* M.S. Thesis, Environmental Toxicology, Utah State University, Logan, UT, 1993.
14. Ferro, A. M.; Sims, R. C; Bugbee, B. G. Plants accelerate the degradation of pentachlorophenol in soil. *Journal of Environmental Quality.* **1993** (in press).
15. Sorensen, D. L.; Watkins, J. W.; Sims, R. C. Plant-enhanced biodegradation of naphthalene. Dept. of Civil and Environmental Engineering, Utah State University, Logan, UT. 1993.
16. Sims, R. C.; Matthews, J. E.; Huling, S. G.; Bledsoe, B. E.; Randolph, M. E.; Pope, D. F. Evaluation of full-scale in situ and ex situ bioremediation of creosote wastes in soils and groundwater. In *Symposium on bioremediation of hazardous wastes: research, development, and field evaluations.* EPA/600/R-93/054. USEPA, Office of Research and Development, 1993.
17. Draft guidance: simple random sampling, systematic random sampling and composting, Research Triangle Institute, Center for Environmental Measurements and Quality Assurance, 1991.
18. USEPA, SW 846 In *Test method for evaluating solid waste physical/chemical methods,* 1986.
19. Page, A. L. (Eds), Cultural methods for soil microorganisms, In *Methods of soil analysis, part 2, chemical and microbiological properties 2nd ed., Agronomy* . Am. Soc. Agron., Madison, Wisconsin, 1982, Vol. 9.
20. Singh, N. B., Singh, P. P., Nair, K. P. P., Effects of legume intercropping on enrichment of soil nitrogen, bacterial activity and productivity of associated maize crop, 1986.
21. Qiu X., HS&ET internal report, Union Carbide Corporation, 1993.

RECEIVED February 22, 1994

PESTICIDES

Chapter 14

Propanil Metabolism by Rhizosphere Microflora

Robert E. Hoagland, Robert M. Zablotowicz, and Martin A. Locke

Southern Weed Science Laboratory, Agricultural Research Service, U.S. Department of Agriculture, P.O. Box 350, Stoneville, MS 38776

Propanil [N-(3,4-dichlorophenyl)propanamide] is a widely-used rice (*Oryza sativa* L.) herbicide. Greenhouse and *in vitro* rhizosphere suspension studies indicate a rapid dissipation of propanil by rice rhizosphere microflora [>90% dissipated in 48 h with up to 70% accumulating as 3,4-dichloroaniline (DCA)]. Aryl acylamidases are responsible for the initial degradation of propanil to DCA and propionic acid. This enzyme is found in a wide range of rhizosphere microflora, including fungal genera such as *Fusarium* and *Trichoderma*, and Gram-negative bacteria genera including *Alcaligines, Bradyrhizobium, Pseudomonas*, and *Rhizobium*. Bacterial strains have been isolated from rice rhizospheres that can completely metabolize 10 μg ml^{-1} propanil to DCA within 30 min. Microbes with potential for rapid metabolism of propanil can be obtained, but their potential for bioremediation may be limited by the formation of more toxic products if appropriate downstream biotransformation mechanisms are lacking.

The use of agrochemicals to provide nutrients to crops and to control weeds, insects, and diseases allows production of adequate quantities of affordable, high-quality food and fiber. Weeds are costly to food production, and herbicide sales approach 50% of the total pesticide sales per year on a global basis (*1*). In the U.S. alone, annual herbicide use in the 1980s was between 173 to 204 million kg (*2*). Propanil [N-(3,4-dichlorophenyl)propanamide] is a herbicide used worldwide for weed control in rice (*Oryza sativa* L.) production. In the two major rice-producing states in the U.S. (AR and LA), 3.4 million kg of propanil were applied to rice fields in 1992 (*3*). Because large quantities of herbicides such as propanil are used, knowledge of the long-term environmental impact of these compounds and their metabolites or degradation products is important. The purpose and scope of this paper is to briefly review some of the pertinent literature on the development, use, toxicology, metabolism, and environmental fate of propanil, and to present some recent findings from our laboratory on propanil metabolism by microflora from the rice rhizosphere. This pertinent information and new data will be discussed in relation to potential problems and oppportunities for bioremediation of propanil and its downstream catabolic products.

Discovery and Use of Propanil

The herbicidal chemistry of propanil (Figure 1) was discovered in 1958 (*4*). Studies of a large variety of acylanilide analogs indicated that maximum phytotoxicity was obtained with 3,4-dichloro- substitution as opposed to monochloro- or other dichloro-analogs (*5*). The primary molecular mode of action of propanil in plants is inhibition of photosynthesis by disruption of e⁻ transport in photosystem II (*6*). This contact herbicide alone is effective in the selective postemergence control of grasses and various broadleaf weeds in rice, and is also formulated with other herbicides such as MCPA, BAS 514, molinate, oxadiazon, 2,4-D, and bromoxynil to broaden the weed control spectrum in some other crops. For example, propanil plus MCPA is used in hard red spring and durum wheat (*Triticum durum* L.) to control *Setaria* spp. grasses and broadleaf weeds (*7*).

Propanil Metabolism

Plants. The first metabolic step in propanil metabolism is enzymatic (aryl acylamidase; EC 3.5.1.a) cleavage of the amide bond to yield 3,4-dichloroaniline (DCA) and propionic acid (Figure 1). Early investigations of propanil metabolism in rice seedlings indicated a rapid conversion to DCA and propionic acid (*8-10*). Later, an aryl acylamidase was isolated and characterized in rice (*11*) and shown to be the biochemical basis of propanil selectivity in the control of barnyardgrass [*Echinochloa crus-galli* (L.) Beauv.] in rice since barnyardgrass tissue could not detoxify absorbed propanil due to very low enzymatic activity. Rice leaves contained a 60-fold greater amount of aryl acylamidase activity than did barnyardgrass leaves. Rice plants also form complexes of DCA, including N-(3,4-dichlorophenyl)glucosylamine (Figure 1) (*12*). The metabolic fate of the propionic acid moiety has also been reported (*9*). Catabolism was found to proceed via β-oxidation with both susceptible and tolerant plant species converting a large amount of the ^{14}C-propanil labeled in either the C-1 or C-3 of the propionate moiety to $^{14}CO_2$. DCA can also be conjugated into lignin complexes in rice (*13*). Aryl acylamidases capable of propanil hydrolysis have been shown to occur in various weed and crop plants (*14, 15*), and some plant aryl acylamidases have been isolated and partially characterized from tulip (*Tulipa gesnariana* v.c Darwin) (*16*), dandelion (*Taraxacum officinale* Weber) (*17*), and red rice (*Oryza sativa* L.) (*18*). Red rice is a serious conspecific weed pest in cultivated rice fields in the southern U.S. (*19*). Recently propanil degradation by aryl acylamidases has been examined in ten wild rice (*Oryza*) species (*20*).

Microbes/Soils/Water. Although propanil may undergo oxidative polymerization

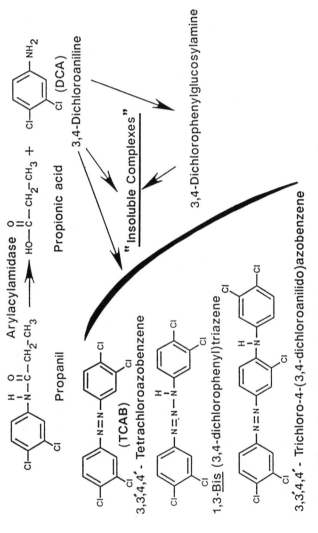

Figure 1.   Schematic of propanil metabolic and/or degradation products.

reactions (*21*), propanil hydrolysis giving rise to DCA (*22, 23*) is the major mechanism for dissipation in soil. The major fraction of the DCA, however, undergoes covalent binding to soil organic matter, supposedly via spontaneous chemical reactions (*24, 25*). This DCA-humus-bound material is immobilized and detoxified since high concentrations do not inhibit microbial respiration (*26*).

DCA can then be catabolized via microbial peroxidases (*27*) to TCAB (*22, 28-31*), TCAOB (*21, 32*), 1,3-bis(3,4-dichlorophenyl)triazene (*33*), and 4-(3,4-dichloroanilido)-3,3′,4,4′-trichloroazobenzene (*34*) (Figure 1). Subsequent dechlorination, ring cleavage, and mineralization to CO_2 of DCA and these other metabolites from propanil in soil is a slow process. Thus there is a persistent nature to some of these metabolites (*35, 36*). However, the persistence and turnover rate of these compounds is dependent on the amount of propanil applied and the soil type (*35*). Polymeric and residual DCA in humus complexes were metabolized to CO_2, and *Aspergillus versicolor* and *Penicillium frequentans* released DCA from these humic complexes (*25*). In flooded rice field water, the applied propanil concentration was reduced rapidly, and only metabolites of propanil were found ~ 15 days after application (*37*).

Many diverse microorganisms have the ability to hydrolyze propanil to DCA (Table I), and enzymes from some have been isolated, partially purified, and characterized. Other microorganisms have been examined for potential metabolism of the major hydrolytic metabolite of propanil, *i.e.*, DCA. As discussed earlier, DCA is resistant to biodegradation in pure and mixed cultures in soils.

E. coli (*54*), *Pseudomonas putida* (*46*), and a mixed microbial culture (*55*) have been shown to have very active metabolic rates on DCA. The most rapid dechlorination of dihalogenated aromatics occurs under reductive/anaerobic conditions (*56, 57*). Under these anaerobic conditions, dehalogenation of the *para* position to form 3-chloroaniline is the major product which occurs rapidly with acclimated cultures. Zeyer and Kerney isolated a pseudomonad (*Pseudomonas* strain G) that was capable of growth on *p*-chloroaniline as a sole carbon source (*58*). The metabolic pathway includes hydroxyl substitution of the amide group and formation of chlorocatechol as the intermediate prior to ring cleavage to form 2-hydroxy-5-chloromuconic semialdehyde (Figure 2). An isolated *P. multivarans* that used aniline as a sole nitrogen and carbon source could also degrade 3-chloroaniline and 4-chloroaniline via chlorocatechol formation to CO_2 in the presence of glucose and aniline (*59, 60*). An *Alcaligenes faecalis* strain could also degrade 3-chloroaniline and 4-choloroaniline in the presence of aniline to 4-chlorocatechol and then to 2-hydroxy-5-chloromuconic semialdehyde (*61, 62*). In both cases, aniline was necessary for this metabolic route because 2,3-catechol oxidases were induced by aniline and not by chloroanilines alone. *Pseudomonas* strain G was capable of mineralizing 3,4-DCA only when supplied with *p*-chloroaniline, but *Moraxella* strain G (*63*) exhibited little or no mineralization of DCA.

TCAB found in soils can arise from contaminated herbicide products (*64, 65*), phototransformation (*66, 67*), and via microbial transformation of DCA (*22, 68-70*). Although propanil degradation in soils is generally quite rapid, the catabolic turnover rates for DCA, TCAB, and TCAOB in soil are relatively slow (*35, 36*) and can vary depending on quantity of propanil applied, soil type, and the ability of the soil to form TCAB and related products (*35*).

Table I. Examples of Microbial Enzymes that Metabolize Propanil to DCA

Organism	Reference
Algae, blue-green	
Anabaena cylindrica	38
Anacystis nidulans	38
Gloeocapsa alpicola	38
Tolypothrix tenuis	38
Algae, green	
Chlamydomonas reinhardii	38
Chlorella vulgaris	38
Bacteria, Gram positive	
Arthrobacter spp.	39
Bacillus sphaericus	40
Norcardia spp.	39
Streptococcus avium	41
Bacteria, Gram negative	
Alcaligenes spp.	42
Pseudomonas spp.	43
P. fluorescens	44
P. pickettii	42, 45
P. putida	41, 46
P. striata	47
Fungi	
Aspergillus nidulans	48
Fusarium oxysporum	49, 50
F. solani	51
Paecilomices varioti	52
Penicillium spp.	53

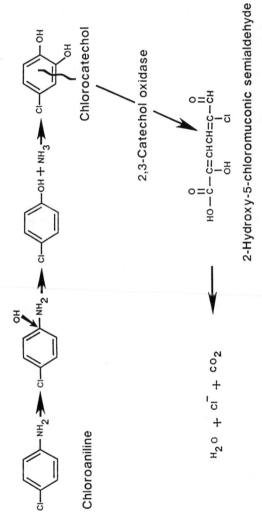

Figure 2. Schematic of chloroaniline metabolism by anaerobic micro-organisms.

Propanil is a photosynthesis inhibitor and could affect oxygen concentration of pond waters by inhibiting phytoplankton photosynthesis, resulting in a deleterious effect on fish (71). Propanil inhibited growth of an alga, *Nostoc calcicola,* and was lethal to the organism at 30 μmol ml^{-1} (72). Various phytoplankton organisms were unable to metabolize propanil (73).

As evidenced by this brief survey of propanil literature, propanil and its products can be transformed in the environment via enzymatic or other degradative pathways. Our experimental objectives were to examine propanil metabolism by rhizosphere microorganisms using greenhouse experiments, *in vitro* studies of rhizosphere suspensions, and *in vitro* studies using enrichment cultures and/or pure cultures of some of these organisms. Details of the methodologies used and our results are presented in the following sections.

Materials and Methods

Chemicals. Propanil (99.0% purity) was purchased from Chem Service, West Chester, PA. Uniformly ring-labeled ^{14}C-propanil (sp. act. = 21.05 μCi/mg) was generously provided by Rohm and Haas Company, Philadelphia, PA.

Cell-free Enzyme Preparation. Fungal cultures were grown for 7 days in potato dextrose broth still-culture at 22° C. Hyphal mats were harvested by filtration, ground in glass homogenizers with K-phosphate buffer (KPi, 0.1 M, pH 6.8), sonicated, centrifuged, and the supernatants filtered (0.2 μm filter). Bacterial cultures were grown for 2 to 3 days in full strength tryptic soy broth (TSB) in shake-cultures (24° C), and cell-free extracts were prepared as for the fungi.

Aryl Acylamidase Assay. Reaction mixtures contained 0.8 μmol propanil and K-phosphate buffer (0.05 M, pH 7.5). Assays were initiated by enzyme addition and incubated at 30° C for 0.5 to 3 h. Termination was achieved with 1N HCl:glacial acetic acid mixture (8:1, v/v). DCA was quantitatively determined using a diazotization procedure (74). Protein concentration was determined with the Bradford reagent (75).

Radiolabeled Propanil Studies. Uniformly ring-labeled ^{14}C-propanil was used in soil, rhizosphere, and whole cell studies. Extracted ^{14}C from various fractions was quantified by liquid scintillation counting (LSC) using Ecolume as cocktail (ICN, Costa Mesa, CA). ^{14}CO$_2$ trapped in mineralization studies was quantified by LSC using Hionic scintillation cocktail.

Thin-Layer Chromatography (TLC). TLC was performed on silica gel (250 μm thick) coated glass with a preabsorbent zone. Plates were developed for 10 cm in benzene:acetone (10:1, v/v), benzene:hexane:ethanol (90:210:30, v/v/v), or benzene (100%). Plates were spotted (25 to 50 μl) using a multi-spotter. Distribution of radioactivity on developed plates was analyzed with an imaging scanner (Bioscan 200). Rf values for unlabeled standards on chromatograms were determined using a Shimadzu CS-9000 dual-wavelength flying-spot scanning densitometer. Reflection at 240 nm was recorded in the zigzag mode. Analysis of standard concentrations indicated that propanil and DCA gave linear responses at 10 to 50 μg ml^{-1} and 5 to 50 μg ml^{-1}, respectively.

Greenhouse Study on Propanil Dissipation in Rice Rhizospheres. Rice seedlings were grown in Conetainers (one seedling per cone) filled with a sandy loam soil (200 g per cone; no herbicide history; pH 6.5), arranged in a randomized complete block of four replicates.   All cones were watered periodically with amounts adequate to maintain soil dampness, yet prevent leaching from the bottom of the cone.   Seven days after planting, propanil was applied as a liquid drench to achieve a soil concentration of 3 μg g^{-1} soil.   Additional watering was not necessary until 48 h after propanil application, when cones were watered sparingly.   At 4, 24, 48, and 96 h after treatment one cone from each block was sacrificed for determination of propanil and DCA.   The contents of each cone (soil and fibrous roots) were homogenized and 50 g of the moist soil-root mixture was extracted twice with 80 ml of MeOH.   Soil moisture was determined on a 20-g subsample, by drying at 85° C for 48 to 72 h. The two MeOH extracts were combined and reduced to 20 ml with a centrifugal evaporator.   The remaining liquid was brought up to 100 ml with distilled water and eluted through a C-18 solid phase extraction (SPE) column.   Propanil metabolites were eluted from the SPE column with 3.0 ml of MeOH.   Propanil and DCA were determined by TLC (benzene:acetone, 10:1, v/v) based upon reflection at 240 nm, as described above.

Bacterial populations of root-free soil and rhizospheres were determined from soil and rice rhizospheres that were untreated or treated with propanil 7 days after treatment.   One entire rice seedling root or 10 g of soil was extracted in 100 ml and 95 ml of KPi (0.1 M, pH 7.0), respectively, shaking for 1.0 h.   Root washings were gently sonicated in a sonicator bath (Branson 5200) for an additional 2 min to aid recovery of rhizosphere microflora, prior to serial dilution.   Total bacterial and Gram-negative bacterial populations were estimated on 10% tryptic soy agar (TSA) and crystal violet media, respectively.   Soil moisture was determined as described above. Weights of soil recovered from rice roots were also determined.   Bacterial populations are reported as log (10) cfu g^{-1} oven-dried soil.

Propanil Degradation by Rice Rhizosphere Bacterial Populations.   Propanil metabolism was studied by total recoverable rhizosphere populations and rhizosphere bacterial enrichments.   Bacterial enrichment cultures were established here by amplification of given species by growth on 10% TSB.   Total rhizosphere populations were recovered from rice roots (14 days after propanil application).   Approximately 8 g of roots were extracted in 100 ml KPi as described above.   After initial extraction, roots were removed and the soil suspension was allowed to stand for 20 min for the coarser soil particles to settle.   Rhizosphere washings were removed and centrifuged at 8,000 xg for 10 min.   The pellet was suspended in 10 ml KPi buffer and plate counts determined as described above.   Propanil was added to 1.0 ml of cell suspension to achieve a concentration of 10 μg ml^{-1} containing 0.2 μCi ml^{-1}. Incubations were conducted in triplicate, and a control of KPi without cells.   Aliquots (100 μl) were removed at 0, 2, 4, 21, 31, 48, 56 and 71 h.   MeOH (100 μl) was added and extraction proceeded via vortexing (30 sec), sonication (2 min), and centrifugation (10,000 xg, 5 min).   The supernatant (25 μl) was spotted on silica plates and developed for 10 cm with benzene:acetone (10:1, v/v). The distribution of radioactivity was monitored by linear imaging scanning. After the final aliquot was taken, total radioactivity remaining in each incubation tube was determined by LSC.

With increased incubation time, lower recovery of readily-extractable ^{14}C material was observed using the above short-term MeOH extraction. Thus, a more intense or long-term sequential extraction was conducted. The remaining cell suspension (250 μl) was extracted with MeOH (400μl) for 1 h, centrifuged, and the pellet extracted using ethyl acetate:KPi (3.5:0.5, v/v; pH 6.8) for 1 h, centrifuged, and the ethyl acetate layer removed. The remaining pellet aqueous layer was extracted with benzene (4.0 ml) for 24 h. Radioactivity was determined from 100 μl aliquots of each sample and the remaining pellet. The ethyl acetate and benzene extracts were reduced to 100 μl, and 40 μl aliquots of all three extracts were spotted on TLC plates and developed in benzene:acetone (10:1) and benzene alone solvent systems. Chromatograms were analyzed for distribution of radioactivity by linear image scanning.

Propanil Metabolism by Rhizosphere Enrichment Cultures. Rhizosphere enrichments from rice roots (7 days after propanil application) were prepared as described elsewhere in this volume (76). Propanil metabolism was determined as described above for radiolabelled propanil except that samples were removed at 2 and 24 h with *Pseudomonas cepacia* strain AMMD included as a reference strain. Aryl acylamidase activity was also determined on cell-free extracts. Nine individual colonies were selected from TSA plates from replicates A and B (high activity). Both aryl acylamidase assays and metabolism of ^{14}C-labeled propanil tests were conducted on cell-free extracts and whole cell preparations.

Mineralization/Metabolism Studies. Small biometer flasks (25 ml) were used to assay for propanil mineralization by cell suspensions of three rhizosphere bacterial strains (AMMD, RA2, and RB4). The flasks were inoculated with 200 μl propanil (50 μg ml^{-1}; 55,000 dpm) and 1.8 ml of bacterial cells (log 10.5 cells ml^{-1}). Flasks were sealed with rubber septum stoppers and fitted with a CO_2 trap containing 400 μl 1N NaOH. Trapped $^{14}CO_2$ was collected at 0, 0.5, 1, 24, 48, 72, and 96 h after treatment, and quantified via LSC in Hionic scintillation cocktail.

Following the 96-h incubation, a modified extraction scheme was implemented. Liquids from all three replicates were combined, centrifuged, and the supernatant concentrated by SPE C-18 columns with MeOH as the extractant. The remaining pellet was sequentially extracted with neutral ethyl acetate (1 h), followed by acidic ethyl acetate (1 h), and finally benzene (24 h). Radioactivity was monitored in all extracts and in the final pellet by LSC. Solvent extracts were concentrated to approximately 100 μl, and 40 μl aliquots were spotted on TLC plates and developed in each of the three solvent systems previously described.

Results and Discussion

Propanil-Aryl Acylamidase Activity in Bacterial and Fungal Reference Cultures.
Hydrolysis of propanil to DCA by the enzyme aryl acylamidase has been reported for several genera of bacteria, including Gram-positive and Gram-negative species, as

presented earlier (Table I). In a first approach to study propanil metabolism, we initiated a survey of propanil-aryl acylamidase (Pro-AAA) activity among isolates of rhizobacteria (typically biological control and/or plant growth-promoting strains) for which taxonomic identification was available. The objective was to ascertain whether certain genera of bacteria may be dominant in propanil degradation in the rhizosphere. Pro-AAA was assayed in a collection of 24 rhizosphere-competent bacteria as summarized in Table II. No Pro-AAA activity was detected in the 7 Gram-positive bacterial strains tested (3 *Arthrobacter* and 4 *Bacillus*). Pro-AAA activity was observed among all strains of *Alcaligines*, *Bradyrhizobium*, and *Rhizobium* tested. Among isolates of the *Pseudomonadaceae*, high activity was observed in the non-fluorescent *P. cepacia* strain AMMD, with a trace of activity found among fluorescent *P. fluorescens* and *P. putida* strains, and no activity detected in the *Xanthomonas campestris* strain. Also, no Pro-AAA activity was detected in any of the six Enterobacteriaceae isolates (*Citrobacter, Enterobacter, Klebsiella,* and *Serratia*).

Table II. Propanil-Aryl Acylamidase Activity Among Reference Strains of Rhizobacteria

Genus	Gram stain response	Strains tested	Specific Activity $nmol\ mg^{-1}protein\ h^{-1}$
Alcaligines	negative	2	1.7 to 2.7
Arthrobacter	positive	3	nd
Bacillus	"	4	nd
Bradyrhizobium	negative	2	0.7 to 1.5
Citrobacter	"	1	nd
Enterobacter	"	2	nd
Klebsiella	"	2	nd
Pseudomonas cepacia	"	1	11.5
P. fluorescens	"	3	trace
P. putida	"	1	trace
Rhizobium	"	1	15.0
Serratia	"	1	nd
Xanthomonas	"	1	nd

Based on previous reports in the literature, Pro-AAA activity was observed in the Gram-positive bacteria *Arthrobacter* and *Bacillus*; but we have not observed any activity in the 3 *Arthrobacter* or 4 *Bacillus* rhizobacterial strains examined in this study. The rhizosphere bacterial population can be dominantly Gram-negative, thus genera such as *Pseudomonas, Alcaligines,* and *Rhizobium* may be dominant contributors to propanil metabolism to DCA in the rhizosphere.

We examined ^{14}C-propanil metabolism in whole cell suspensions of AMMD and other rhizobacterial strains. ^{14}C-Propanil was not autohydrolyzed in aqueous suspension over a 24-h period. However, strain AMMD metabolized ^{14}C-propanil after 24 h, yielding ^{14}C-DCA and two other major labeled peaks at the origin and Rf 0.18. Negligible hydrolysis of propanil was observed during this time period by other whole cell suspensions of *Pseudomonas* strains listed in Table II (data not shown).

Propanil hydrolysis by several species of *Fusarium,* namely *F. solani* (*51*) and *F. oxysporum* (*49, 50*) has been reported. A collection of *Fusarium* species was available at our laboratory. This study was conducted to assess the contribution of this dominant genus of rhizosphere-competent fungi for Pro-AAA activity in the rhizosphere. Pro-AAA activity was observed among isolates of all five species of *Fusarium* studied (Table III). Although a wide range of activity has been observed among isolates within each species, *F. solani* had more active strains than the other species studied. Pro-AAA activity was also observed in two other fungi studied, *Trichoderma harganinum* and *Phenacrycete chryosporium*. *F. solani* is the dominant *Fusarium* in many soils and rhizospheres of crop plants (*77*). The fusaria may contribute to propanil metabolism under well-drained soil conditions, however under the flooded soil conditions encountered in rice culture their populations rapidly decline (*78*). Other microbes can rapidly metabolize propanil to DCA in anaerobic soils (*79*). Propanil hydrolysis under anaerobic conditons was slightly faster when nitrate-reducing conditions existed, while DCA was reductively dehalogenated, but not mineralized, in methanogenic soil slurries (*80*).

Table III. Propanil Aryl Acylamidase Activity Among *Fusarium* Isolates and Other Fungal Species

Genus species	Number of strains tested	Specific Activity, nmol DCA mg^{-1} protein h^{-1} (range / # > 1.0)
Fusarium equisiti	2	nd - 1.3 / 1
F. moniliforme	6	nd - 13.8 / 2
F. oxysporum	6	nd - 4.6 / 2
F. semitectum	2	2.5 - 4.7 / 2
F. solani	10	nd - 8.2 / 8
Trichoderma harzaninum	1	0.5
Phenacrycete chryosporium	1	0.1

Greenhouse Studies on Propanil Dissipation in Rice Rhizospheres. Studies of propanil in rice rhizospheres indicated that after 4 h, only 65% of the initial propanil was found in the MeOH-soluble fraction (Figure 3), and DCA was observed in all samples within hours after application. MeOH-extractable propanil declined rapidly over the 4-d time course. MeOH-extractable DCA increased from 4 h to 2 d and then declined. Since non-labeled propanil was used in these studies, we were limited in ability to further evaluate metabolism of propanil beyond DCA. However, it is evident that a major fraction of propanil metabolites are bound to soil, mineral, or organic colloids, as previously reported (*24, 25*).

Analysis of bacterial populations after propanil application indicated little or no effect on bacteria in soil, but there was a slight effect on reducing the rhizosphere Gram-negative bacterial population (Table IV). Gram-negative bacterial populations were enriched 30-fold in the rhizosphere soil, and the total bacterial populations were enriched 11-fold in the rhizosphere soil compared to the bulk soil.

Table IV. Effect of Propanil on Soil and Rhizosphere Bacterial Populations and Rhizosphere Enrichment[a]

	Log cfu[b]/g soil		R:S ratio	
Source	Gram neg.	Total	Gram neg.	Total
Rhizosphere				
Propanil	6.88	8.35	26.3	13.41
Nontreated	7.21	8.46	33.1	10.2
Pr[c] > F	0.075	0.465	0.223	0.202
Soil				
Propanil	5.47	7.34	na	na
Nontreated	5.69	7.32	na	na
Pr > F	0.496	0.557		

[a] Mean of four samples.
[b] Colony-forming units = cfu.
[c] Probability of data value being greater than F due to treatment effect (F-test) .

Propanil Degradation by Rice Rhizophsere Bacterial Poplulations. Our second approach to characterize propanil metabolism in the rhizosphere was to extract native rhizosphere populations from rice roots previously exposed to propanil. Microbial populations consisted of log 7.8 (total bacteria), log 6.7 (Gram-negative bacteria), log 4.3 fluorescent pseudomonads and log 4.4 total fungi cfu ml[-1].

During the first 21 h, MeOH-extractable propanil remained relatively constant and thereafter declined rapidly (Figure 4A). DCA was first detected at the 21-h

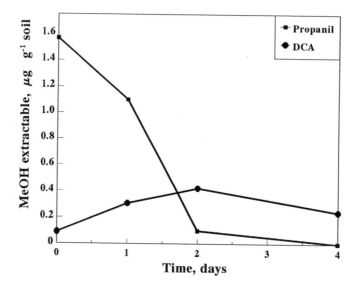

Figure 3.   Propanil dissipation in rice rhizospheres.

sample and continued to increase until 48 h and declined thereafter (Figure 4B). At least two additional metabolites were first detected at 48 h and continued to accumulate until termination, one with an Rf of 0.18 and one (or more) remained at the origin (immobile in the benzene:acetone solvent). This data was based upon a rapid MeOH extraction technique similar to that used in other bacterial propanil metabolism studies (46). By 48 and 70 h approximately 32 and 8% of the initial ^{14}C added was recovered, respectively, indicating sorption of propanil and or metabolites to cellular components.

At the termination of the study a more rigorous extraction methodology was implemented (Table V). Longer duration exposure (1 h) to 50% MeOH resulted in an 83% recovery versus 8% as described above, while less than 13% was recovered in the two subsequent extractions using ethyl acetate and benzene. These results demonstrate that approximately 95% of the propanil was metabolized to DCA and at least five additional products. About 3% of the applied ^{14}C was unextractable and a total recovery of applied ^{14}C was 98.3% indicating that at most, less than 2% was mineralized to $^{14}CO_2$ during this study. Generally, the ethyl acetate extraction was more efficient in the extraction of four metabolites with Rfs of 0.18, 0.45, 0.81, and 0.95. The two nonpolar metabolites (Rf 0.81 and 0.95) may be TCABs since these compounds have similar Rfs in the hexane:benzene:ethanol system as reported for TCABs in the literature (81) and in 100% benzene as a solvent system (82). These two metabolites represent about 4% of the propanil initially added, and in our studies both compounds were either present or absent. We have not, as yet, identified the remaining unknown metabolites. One possible metabolite, *p*-chloroaniline, could be masked since it migrates with DCA in all three TLC solvent systems used in our studies. The TLC profiles observed for these studies are quite similar to profiles observed in studies with *P. cepacia* strain AMMD.

Propanil Metabolsim in Rhizosphere Enrichment Cultures. The third approach we used to study propanil metabolism by rice rhizospheres was the use of rhizosphere enrichment cultures (83). Four rhizosphere enrichment cultures, with bacterial populations of log 10.3 to 10.6 cfu ml^{-1} (47 to 76% Gram-negative) were obtained from rice roots exposed to propanil (Table VI). Pro-AAA activity was observed in cell-free extracts of all four enrichments. However, two mixed cultures, RA and RB, exhibited Pro-AAA activity in crude enzyme preparations ranging from 3 to 9 times greater than we have previously encountered in our laboratory. Radiological assays on whole cells of these mixed cultures with ^{14}C-propanil indicated that all propanil was converted to DCA within 2 h by both RA and RB while less than 10% of the added propanil was hydrolyzed to DCA by cultures RC and RD within a 24-h incubation.

Nine dominant colony types were isolated from mixed cultures RA and RB and purified to single strains (RA1-4 & RB1-5). These were again characterized by both Pro-AAA activity in cell-free extract and ^{14}C-propanil metabolism in whole cell studies. Although Pro-AAA activity was observed in 6 of the 9 strains isolated, two isolates, RA2 and RB4, were the organisms predominantly responsible for high the Pro-AAA activity in the original mixed cultures (Table VI).

Preliminary taxonomic analysis of strains RA2 and RB4 indicates that both are Gram-negative rods with multiple polar flagella. Further taxonomic classification/

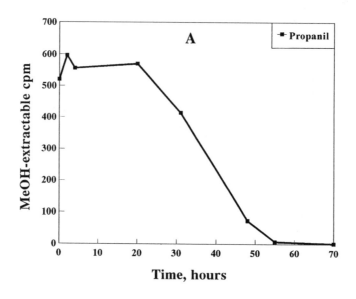

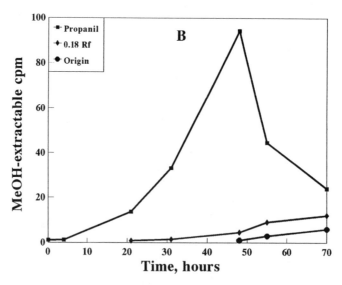

Figure 4.   Propanil metabolism in rhizosphere suspensions.   A. Propanil
dissipation.   B. Metabolite profile.

Table V. Recovery of ^{14}C from Propanil-Treated Rhizosphere Suspensions in Sequential Extraction Solvents, 71 h After Treatment

Extraction	$\%\,^{14}C$ Applied	Propanil 0.34	DCA 0.58	$\%\,^{14}C$ Recovered Rf (Benzene:Acetone, 10:1, v/v)				
				Origin	0.18	0.45	0.81	0.95
Methanol	82.6 ± 5.6	3.5 ± 3.7	70.3 ± 2.6	13.6 ± 3.8	9.1 ± 0.9	nd	0.8 ± 1.3	2.7 ± 2.6
Ethyl Acetate (neutral)	7.8 ± 1.3	8.8 ± 5.5	65.2 ± 12.6	14.3 ± 2.3	15.2 ± 6.0	2.4 ± 2.6	3.8 ± 3.3	8.9 ± 8.5
Benzene	5.1 ± 2.6	15.4 ± 5.2	60.0 ± 6.0	12.4 ± 3.2	7.7 ± 1.0	nd	6.5 ± 2.3	3.7 ± 6.4
Unextractable	2.8 ± 0.7	-	-	-	-	-	-	-
Total ^{14}C recovered, %	98.3	4.3	66.0	12.7	12.5	0.2	1.3	3.0

identification of these two strains is in progress. The specific activity of strains RA2 and RB4 are several fold greater than that previously reported for crude enzyme preparations of soil *Alcaligines* and *Pseudomonas* isolates (*45*).

Table VI. Propanil-Aryl Acylamidase Activity of Rice Rhizosphere Enrichment Cultures and Single Colony Isolates From These Enrichments

Culture or Isolate	Specific Activity, nmol DCA, mg^{-1} protein h^{-1}
Enrichment cultures	
RA	134.0
RB	45.6
RC	0.6
RD	3.4
Single colony isolates	
RA1	2.0
RA2	2,250 - 15,320
RA3	8.4
RA4	nd[a]
RB1	nd
RB2	nd
RB3	3.0
RB4	3,980 - 12,360
RB5	3.1

[a] None detected (<0.05).

Pure Culture Propanil Metabolism Studies. Two additional ^{14}C-propanil metabolism studies were conducted with *P. cepacia* strain AMMD and isolates RA2 and RB4. The first study was conducted for 24 h and the second study for 96 h. The potential for mineralization of ^{14}C-propanil to CO_2 was evaluated in these studies.

In the first study, $> 98\%$ of readily extractable ^{14}C was recovered as DCA within 30 min by RA2 and RB4, while DCA (5%) was observed by AMMD after 2-h incubation (data not shown). After 24 h, the study was terminated by extraction with 50% MeOH (1 h) followed by a neutral ethyl acetate extraction (1 h). Approximately 85% of the ^{14}C was recovered in the first MeOH extraction and 8 to 10% in the ethyl acetate extraction in all three strains (data not shown). Propanil was observed in both extracts from AMMD (43 and 31%), but only small amounts of propanil (5%) were observed in the ethyl acetate fractions from RA2 and RB4 (Table VII). The polar metabolite (Rf 0.18) was the dominant metabolite formed by AMMD, while 22 to 41% of the radioactivity from all three strains remained at the origin in the MeOH extracts. AMMD likewise accumulated greater nonpolar metabolites (potential TCABs, Rf 0.85 and 0.95) than did RA2 or RB4.

The same three bacterial strains (RA2, RB4, and AMMD) were examined over

Table VII. ^{14}C -Propanil Metabolites by AMMD, RA2, and RB4 in Two Studies

	% Recovery, ^{14}C in Propanil and Metabolites		Rf (Benzene:Acetone; 10:1, v/v)					
	Propanil	DCA	Origin	0.07	0.18	0.47	0.85	0.95
AMMD, 24 h								
50% MeOH	43	2	24	nd	31	nd	nd	nd
EtOAC (neutral)	31	9	17	nd	11	3	5	24
AMMD, 96 h								
KPi	26	nd	9	14	55	nd	nd	nd
EtOAC (Neutral)	10	5	16	6	20	2	nd	21
RA2, 24 h								
50% MeOH	nd	78	22	nd	nd	nd	nd	nd
EtOAC (neutral)	5	74	7	3	8	nd	6	nd
RA2, 96 h								
KPi	nd	100	nd	nd	nd	nd	nd	nd
EtOAC (neutral)	nd	71	12	5	3	5	nd	3
RB4, 24 h								
50% MeOH	nd	53	41	nd	nd	nd	6	nd
EtOAC (neutral)	5	71	5	nd	8	nd	5	6
RB4, 96 h								
KPi	nd	81	11	nd	5	nd	nd	2
EtOAC (neutral)	4	80	7	nd	4	3	nd	2

a long time course (96 h) for propanil metabolism/mineralization in whole cell suspension studies. After 96 h, the aqueous or buffer-soluble fraction from each culture contained the largest amount of extractable radioactivity, but the range of the aqueous-soluble ^{14}C differed; i.e., RA2 = 86%, RB4 = 79%, and AMMD = 65%. Subsequent organic solvent extractions (neutral ethyl acetate, acidic ethyl acetate, and benzene) contained lesser amounts of ^{14}C, but only the neutral ethyl acetate fraction differed among cultures. Unextractable or residual ^{14}C was ~ 1% for RB4 and significantly less in the other two cultures.

Total $^{14}CO_2$ evolved during the 96 h study was low; i.e., ~1.5% for RB4 and AMMD, and only 0.4% for RA2. Less than 0.05% volatilization of propanil was captured in the NaOH trap in the uninoculated controls. Total ^{14}C-recovery was very high (> 98%) for strains RA2 and RB4, but only ~75% for AMMD. During concentration of the KPi extractable fraction, ~ 95% of applied ^{14}C was retained by the C-18 SPE column for strains RA2, RB4 and noninoculated control, while only ~ 63% of the ^{14}C was retained for AMMD. This indicates that a high percentage of the propanil originally applied was converted to highly polar metabolites by strain AMMD. It is also possible that $^{14}CO_2$ may have been underestimated in the AMMD culture because of NaOH-trap saturation.

As observed in previous studies, radiochromatographic scans of TLC plates of recovered ^{14}C-propanil and ^{14}C-metabolites from KPi and neutral ethyl acetate fractions showed that a significant portion of initially applied propanil was recovered only in AMMD (26% in KPi-soluble fraction and 10% in the neutral ethyl acetate extract). Again the Rf 0.18 metabolite was the dominant ^{14}C-labeled product from AMMD, and DCA was the major product observed in both RA2 and RB4. The non-polar metabolite at Rf 0.95 was not observed in any of the KPi-soluble fractions but was one of the major neutral ethyl acetate-soluble products from AMMD, while a lesser amount (2 to 3%) was observed in both RA-2 and RB4. We followed the extraction protocol previously described by Kearney and Plimmer (81) to attempt to find products of ring cleavage. Very little of the radioactivity was observed in the acidified ethyl acetate fraction which should have contained the catechol and 2-chloromucuronic semialdehyde, if present (≤ 1%).

The accumulation of DCA by strain AMMD varied in the three whole-cell metabolism studies. In the first study, DCA accounted for 25% of the ^{14}C recovered, while in these two studies most DCA was already metabolized to other products. Some of the polar metabolites, i.e., materials remaining at the origin and Rf 0.18, may be conjugated products, a strategy found in many organisms to detoxify DCA. Two patterns of propanil metabolism have been observed in these three strains. Strains RA2 and RB4 rapidly hydrolyze propanil to DCA with very little further metabolism. Metabolically, strain AMMD is more interesting because the DCA that is formed is readily metabolized into several other compounds, including possible TCABs. As described previously in the rhizosphere suspension studies, the AMMD metabolite (Rf 0.95, benzene:acetone, 10:1) was verified in two additional solvents, hexane:benzene:ethanol (90:210:30) and benzene (100%) (data not shown) with Rfs similar to those reported for TCAB (81, 82). Similar Rfs were reported for TCAB (data not shown). Purification and identification of some of these metabolites is currently in progress.

Conclusions/Potential for Bioremediation

Progress has been made in the past few years in the knowledge of microbial degradation mechanisms of halogenated aromatic molecules. There is recent and considerable interest in developing methods for microbiological *in situ* clean-up of water and soil contaminated with organic halogen compounds, as evidenced in reviews *(83, 84)*. Inoculation of soils with selected bacterial cultures to detoxify specific pesticides has been somewhat successful. For example, pentachlorophenol residues have been decreased by soil amendments of an *Arthrobacter* sp. *(85)*. Success was also demonstrated with a pentachlorophenol-degrading *Flavobacterium* sp., but several innoculations were required for the removal of PCP from soils *(86)*. Also, a 2,4,5-T-degrading *P. cepacia* strain was useful in decontamination of high 2,4,5-T residues in soils *(87)*. Studies on pesticide metabolism in pure cultures of microbes play an important role in the understanding of biodegradation, and in some instances have led to the isolation and characterization of specific degradation enzymes. Pure culture metabolic studies can aid in the understanding of processes relating to regulation of metabolic pathways and lead to an understanding of expression of the active enzymes under various physical-chemical interactions.

Several factors make bioremediation with microbes a very complex process. For successful bioremediation, the amendment organism must be able to adapt and compete with the existing native microflora. Therefore, the ecology, physiology, and biochemical degradative mechanisms must be adequate to cope with the specific environmental conditions in a given clean-up situation and perhaps to tolerate other agrochemicals in the soil. As pointed out earlier, the complete metabolism of propanil presents some problems since several down-stream metabolites (DCA, TCAB, TCAOB, and others) are more persistent and toxicologically more harmful than propanil itself. However, it is also noteworthy to point out that compounds from other herbicide classes, such as the ureas and carbamates, can also degrade to DCA. There are several potential pathways for conjugation of DCA; e.g. acetylation and glucosylation that can detoxify DCA. These metabolic pathways can minimize the formation of some of the more toxic diazo compounds. Furthermore, the fact that DCA and some of its metabolites are bound/complexed with organic or humic acid soil fractions has obscured the real toxicological significance and impact of these compounds *(88)*.

As pointed out in earlier literature and in our work presented here, initial propanil metabolism via aryl acylamidases is quite widespread among bacteria and fungi. Our strategy, however, is somewhat unique in that we examined rhizosphere microorganisms rather than total bulk soil microorganisms. Three approaches have been taken to study propanil metabolism in the rhizosphere (greenhouse study, *in vitro* studies with native rhizosphere suspensions, and *in vitro* rhizosphere enrichment experiments). Data from these three approaches suggest that bacterial metabolism in the rhizosphere may be a dominant factor in propanil dissipation in rice-production soils.

We have observed an enrichment of bacteria, especially Gram-negative bacteria, in the rhizosphere of rice, as reported elsewhere *(78, 89, 90)*. Our taxonomic studies, as well as the work of others, indicate that several genera of Gram-negative bacteria (especially *Alcaligines*, *Pseudomonas*, and *Rhizobium*) are capable of rhizosphere competence and possess the enzymatic potential for propanil degradation. Initial selection procedures have obtained rhizobacteria (RA2 and RB4) with potential to rapidly transform propanil to DCA. Further work will be necessary to select the most valuable rhizosphere organisms for further study for bioremediation. However, if such rhizosphere organisms are found, they may inherently be more competent and have higher survival potential due to their association with plant roots and plant exudates. Their proximity to roots and root exudates would also enable rapid metabolism of translocated residues from foliar-applied agrochemicals and/or prevent crop plant uptake of phytotoxic or toxicological residues existing in soils. As mentioned above, bioremediation organisms will have to tolerate the presence of naturally-occurring chemicals and agrochemcials. This could pose a problem for the first step in propanil metabolism, since aryl acylamidases are known to be inhibited by organophosphorus and carbamate insecticides (*11, 91, 92*).

Although strain AMMD exhibits lower aryl acylamidase activity than our native rhizosphere community, propanil degradation profiles were similar. Strain AMMD may be valuable for pure culture simulations of propanil dissipation in the rhizosphere environment.

Propanil (3 to 5 kg ha^{-1}) is typically applied post-emergence and usually 1 to 2 days prior to flooding, depending upon the degree of barnyardgrass infestation. Although the rice soils will remain flooded during the remainder of the season, aerobic metabolism in the rhizosphere will still be dominant because the rice root system provides for an oxygen rich environment under this otherwise reduced environment (*93*). Studies on biological control of fungi using bacterial inoculants have indicated that certain pseudomonads can be introduced to the rice rhizosphere and maintain satisfactory populations despite the flooded conditions (*94*). Enhanced propanil degradation via modification of the rice rhizosphere may be a worthy effort for future research.

Acknowledgements

We thank R.E. Gordon and M.E. Smyly for their excellent technical contributions. We are very appreciative to: Charles Hagedorn, VA Polytechnic Institute and State University; Charles Howell, USDA-ARS, College Station, TX; Mark Mount, Univ. Massachusettes, Amherst; John M. Kloepper and Joe Kloepper, Auburn Univ.; Jennifer Parke, Univ. Wisconsin; Stan Nemec, USDA-ARS, Orlando, FL; Eugene Milus, Univ. Arkansas. We thank Hamad Abbas for contributions of fungal cultures and assistance in their preparation for enzyme analysis. We also wish to thank S.O. Duke, Khrishna Reddy, R.D. Williams, and G.D. Wills for their critical reviews of this manuscript.

Literature Cited

1.   Jutsum, A.R. *Phil. Trans. Royal Soc. London B* **1988**, *318*, 357-73.
2.   National Research Council, *Alternative Agriculture*; National Acad. Press; **1989**, 448 pp.
3.   Crop Production; 1992 Summary and Agricultural Usage Survey. National Agricultural Statistics Service, USDA.

4. W. Schäfer et al., Ger. pat. #1,039,779, **1958**.
5. Huffman, C.W.; Allen, S.E. *J. Agric. Food Chem.* **1960**, *8*, 298-302.
6. Hofstra, G.; Switzer, C.M. *Weed Sci.* **1968**, *16*, 23-8.
7. *Weed Sci. Soc. Amer. Herbicide Handbook*, 6th Ed. Weed Sci. Soc. Amer.; Champaign, IL, **1989**, p 220-1.
8. Adachi, M.; Tonegawa, K.; Ueshima, T. *Noyaku Seisen Gijutsu*, **1966**, *14*, 19-22.
9. Still, G.G. *Plant Physiol.* **1968**, *43*, 543-6.
10. Unger, J.H.; McRae, D.H.; Wilson, H.F. *WSSA Abstracts* **1964**, p. 86.
11. Frear, D.S.; Still, G.G. *Phytochemistry*, **1968**, *7*, 913-20.
12. Still, G.G. *Science*, **1967**, *159*, 992-3.
13. Yih, R.Y.; McRae, D.H.; Wilson, H.F. *Science*, **1968**, *161*, 376-7.
14. Hoagland, R.E.; Graf, G. *Weed Sci.* **1972**, *20*, 303-5.
15. Hoagland, R.E.; Graf, G.; Handel, E.D. *Weed Res.* **1974**, *14*, 371-4.
16. Hoagland, R.E.; Graf, G. *Phytochemistry*, **1972**, *11*, 521-7.
17. Hoagland, R.E. *Phytochemistry*, **1975**, *14*, 383-6.
18. Hoagland, R.E. *Plant Cell Physiol.* **1978**, *19*, 1019-27.
19. Smith, Jr., R.J.; Flinchum, W.T.; Searman, D.E. *U.S. Dept. of Agric., Agric. Handbook No. 497*, **1977**, 78 pp.
20. Jun, D.J.; Matsunaka, S. *Pestic. Biochem. Physiol.* **1990**, *38*, 26-33.
21. Bartha, R.; Pramer, D. *Adv. Appl. Microbiol.* **1970**, *13*, 317-41.
22. Bartha, R.; Pramer, D. *Science*, **1967**, *156*, 1617-8.
23. Bartha, R. *J. Agric. Food Chem.* **1968**, *16*, 602-4.
24. Bartha, R. *J. Agric. Food Chem.* **1971**, *19*, 385-7.
25. Hsu, T.-S.; Bartha, R. *J. Agric. Food Chem.* **1976**, *24*, 118-22.
26. Hsu, T.-S.; Bartha, R. *Soil Sci.* **1974**, *118*, 213-20.
27. Pothuluri, J.V.; Hinson, J.A.; Cerniglia, C.E. *J. Environ. Qual.* **1991**, *20*, 330-47.
28. Bartha, R.; Linke, H.A.B.; Pramer, D. *Science*, **1968**, *161*, 582-3.
29. Belasco, I.J.; Pease, H.L. *J. Agric. Food Chem.* **1969**, *17*, 1414-7.
30. Sprott, G.D.; Corke, C.T. *Can. J. Microbiol.* **1971**, *17*, 235-40.
31. Cripps, R.E.; Roberts, R.T. In *Pesticide Microbiology*; Hill, J.R.; Wright, S.J.L.; Eds.; Academic Press: New York, **1978**, pp. 669-730.
32. Kaufman, D.D.; Plimmer, J.R.; Iwan, J.; Klingebiel, U.I. *J. Agric. Food Chem.* **1972**, *20*, 916-9.
33. Plimmer, J.R.; Kearney, P.C.; Chisaka, H.; Yount, J.B.; Klingebiel, U.I. *J. Agric. Food Chem.* **1970**, *18*, 859-61.
34. Linke, H.A.B,; Bartha, R. *Naturwissenschaften*, **1970**, *57*, 307-8.
35. Chisaka, A.H.; Kearney, P.C. *J. Agric. Food Chem.* **1970**, *18*, 854-8.
36. Lee, J.K.; Fournier, J.C. *J. Korean Agric. Chem. Soc.* **1978**, *21*, 71-80.
37. Popova, G.V. *Eksp. Vodn. Toksikol.* **1973**, *4*, 38-49.
38. Wright, S.J.L.; Maule, A. *Pestic. Sci.* **1982**, *13*, 253-6.
39. Burge, W.D. *Soil. Biol. Biochem.* **1972**, *4*, 378-86.
40. Engelhardt, G.; Wallnöfer, P.R.; Plapp, R. *Appl. Microbiol.* **1973**, *26*, 709-18.
41. Dahchour, A.; Britton, G.; Coste, C.M.; Bastide, J. *Bull. Environ. Contam. Toxicol.* **1986**, *36*, 556-62.

42. Hirase, K.; Matsunaka, S. *Pestic. Biochem. Physiol.* **1991**, *39*, 302-8.
43. Zeyer, J.; Kearney, P.C. *Pestic. Biochem. Physiol.* **1982**, *17*, 224-31.
44. Hammond, P.M.; Price, C.P.; Scawen, M.D. *Eur. J. Biochem.* **1983**, *132*, 651-5.
45. Hirase, K.; Matsunaka, S. *Proc. Br. Crop Prot. Conf. Weeds*, **1989**, *2*, 419-26.
46. You, I.-S.; Bartha, R. *J. Agric. Food Chem.* **1982**, *30*, 274-7.
47. Kearney, P.C. *J. Agric. Food Chem.* **1965**, *13*, 561-4.
48. Pelsy, F.; Leroux, P.; Heslot, H. *Pestic. Biochem. Physiol.* **1987**, *27*, 182-8.
49. Blake, J.; Kaufman, D.D. *Pestic. Biochem. Physiol.* **1975**, *5*, 305-13.
50. Reichel, H.; Sisler, H.D.; Kaufman, D.D. *Pestic. Biochem. Physiol.* **1991**, *39*, 240-50.
51. Lanzilotta, R.P.; Pramer, D. *Appl. Microbiol.* **1970**, *19*, 301-6.
52. Hiramatsu, A.; Yasumoto, S.; Kodama, O.; Akatsuka, T. *Agric. Biol. Chem.* **1982**, *46*, 1751-6.
53. Sharabi, N.E.-D.; Bordeleau, L.M. *Appl. Microbiol.* **1969**, *18*, 369-75.
54. Bunce, N.J.; Merrick, R.L.; Corke, C.T. *J. Agric. Food Chem.* **1983**, *31*, 1071-5.
55. Surovtseva, E.G.; Vasileva, G.K.; Volnova, A.I. *Mikrobiologija* **1984**, *53*, 10-5.
56. Stepp, T.D.; Caper, N.D.; Paynter, M.J.B. *Pestic. Biochem. Physiol.* **1985**, *23*, 256-60.
57. Struss, J.; Rogers, J.E.M. *A.E.M.* **1989**, *55*, 2527-31.
58. Zeyer, J.; Kearney, P.C. *Pestic. Biochem. Physiol.* **1982**, *17*, 215-23.
59. Reber, H.; Helm, V.; Karanth, N.G.K. *Eur. J. Appl. Microbiol. Biotechnol.* **1979**, *7*, 181-9.
60. Helm, V.; Reber, H. *Eur. J. Appl. Micribiol. Biotechnol.* **1979**, *7*, 191-9.
61. Surovtseva, E.G.; Volnova, A.I.; Shatskaya, T.Y. *Mikrobiologija*, **1980**, *49*, 351-4.
62. Surovtseva, E.G.; Vasileva, G.K.; Volnova, A.I.; Baskunov, B.P. *Dokl. Akad. Nauk USSR*, **1980**, *254*, 226-30 (*Chem Abst. 94*, 59181d).
63. Zeyer, J.; Wasserfallen, A.; Tummis, K.N. *Appl. Environ. Microbiol.* **1985**, *50*, 447-53.
64. Hill, R.H.; Rollen, R.Z.; Kimbrough, R.D.; Groce, D.F.; Needham, L.L. *Arch. Environ. Health*, **1981**, *36*, 11-4.
65. DiMuccio, A.; Camoni, I.; Dommarco, R. *Ecotoxicol. Environ. Saf.* **1984**, *8*, 511-2.
66. Moilanen, K.W.; Crosby, D.G. *J. Agric. Food Chem.* **1972**, *20*, 950-3.
67. Wolff, C.J.M; Crossland, N.O. *Environ. Toxicol. Chem.* **1985**, *4*, 481-7.
68. Kuhr, R.J.; Casida, J.E. *J. Agric. Food Chem.* **1967**, *15*, 814-24.
69. Bordeleau, L.M.; Rosen, J.D.; Bartha, R. *J. Agric. Food Chem.* **1972**, *20*, 573-8.
70. Hughes, A.F.; Corke, C.T. *Can. J. Microbiol.* **1974**, *20*, 35-9.
71. Tucker, C.S. *Bull. Environ. Toxicol.* **1987**, *39*, 245-50.
72. Pandey, A.K.; Srivastava, V.; Tiware, D.N. *Alg. Mikrobiol.* **1984**, *24*, 369-76.

73.   Kuiper, J.; Hanstveit, A.O. *Ecotoxicol. Environ. Saf.* **1984**, *8*, 34-53.
74.   Bratton, A.C.; Marshall, Jr., E.K. *J. Biol. Chem.* **1939**, *128*, 537-50.
75.   Bradford, M. *Anal. Biochem.* **1976**, *72*, 248-54.
76.   Zablotowicz, R.M.; Hoagland, R.E.; Locke, M.A.; (this volume).
77.   Nemec, S.; Zablotowicz, R.M.; Chandler, J.L. *Phytophylactica* **1989**, *21*, 141-6.
78.   Mitchell, R.; Alexander, M. *Soil Sci.* **1962**, *87*, 413-9.
79.   Pettigrew, C.A.; Paynter, M.J.B.; Camper, C.D. *Soil Biol. Biochem.* **1985**, *17*, 815-8.
80.   Suflita, J.M.; Ramanand, K.; Adrian, N. In *Organic Substances and Sediments in Water*; Baker, R., Ed.; Lewis Publishers: Chelsea, MI, **1991**, Vol. 3; pp 199-210.
81.   Kearney, P.C.; Plimmer, J.R. *J. Agric. Food Chem.* **1972**, *20*, 584-5.
82.   Bartha, R. *J. Agric. Food Chem.* **1972**, *19*, 385-7.
83.   Morgan, P.; Watkinson, R.J. *FEMS Microbiol.* **1989**, *63*, 277-300.
84.   Boyle, M. *J. Environ. Qual.* **1989**, *18*, 395-402.
85.   Edgehill, R.U.; Finn, R.K. *Appl. Environ. Microbiol.* **1983**, *45*, 1122-5.
86.   Crawford, R.L.; Mohn, W.W. *Enzyme Microbiol. Technol.* **1985**, *7*, 619-20.
87.   Kilbane, J.J.; Chattergee, D.K.; Chakrabary, A.M. *Appl. Environ. Microbiol.* **1983**, *45*, 1697-700.
88.   Marco, G.J.; Novak, R.A. *J. Agric. Food Chem.* **1991**, *39*, 2101-11.
89.   Holding, A.J. *J. Appl. Bact.* **1960**, *23*, 515-25.
90.   Rovira, A.D.; Davey, C.B. Biology of the Rhizoshpere. In *The Plant Root and Its Environment*; Carson, E.W., Ed.; Univ. Press of Virginia: Charlottesville, **1977**, pp. 153-204.
91.   Bowling, C.C.; Hudgins, H.R. *Weeds*, **1966**, *14*, 94-5.
92.   Matsunaka, S. *Science*, **1968**, *160*, 1360-1.
93.   Armstrong, W. *Physiol. Plant.* **1971**, *24*, 242-7.
94.   Sakthivel, N.; Sivamani, E.; Unnamalai, N.; Gnanamanickam, S.S. *Current Sci.* **1986**, *55*, 22-5.

RECEIVED March 31, 1994

Chapter 15

Glutathione S-Transferase Activity in Rhizosphere Bacteria and the Potential for Herbicide Detoxification

Robert M. Zablotowicz, Robert E. Hoagland, and Martin A. Locke

Southern Weed Science Laboratory, Agricultural Research Center, U.S. Department of Agriculture, P.O. Box 350, Stoneville, MS 38776

Glutathione S-transferase (GST; EC 2.5.1.18) activity was found in isolates of Gram-negative rhizobacteria, especially *Pseudomonadaceae* and *Enterobacteriaceae*. GST-mediated dechlorination of chlorodinitrobenzene and the herbicide alachlor and GST-mediated cleavage of the ether bond of the herbicide fluorodifen was observed for some strains. No atrazine-GST activity was found. Certain fluorescent pseudomonads can dechlorinate 75 to 100 µM alachlor in 48 to 96 h, with the cysteine conjugate as the major accumulating intermediate. Levels of alachlor-GST catabolism observed in native rhizosphere communities were lower than most *Pseudomonas fluorescens* strains identified in this work. Activities of γ-glutamyltranspeptidase (EC 2.3.2.2), and cysteine-β-lyase (EC 4.4.1.6 and 8) enable a rapid metabolism of glutathione-conjugates. Rhizosphere-competent bacteria selected for high GST activity and downstream catabolism may have potential as seed/soil inoculants in bioremediation/detoxification of certain herbicides or other xenobiotics.

The interface between plant roots and soil (the rhizosphere) is a unique niche for the proliferation and selective enrichment of various microbial populations (*1*). This enrichment, due to exudation of carbohydrates, amino acids, and other carbonaceous materials by plant roots, serves as a rich substrate for microbial activity (*1,2*). The concentrations of these substrates are several orders of magnitude greater than those observed in bulk soil. Accelerated degradation of certain pesticides and xenobiotics (e.g. two organophosphorous insecticides parathion (*3,4*), diazinon (*3*)) and trichloroethylene (*5*), has been observed in the rhizosphere compared to bulk soil.

Glutathione S-transferase (GST) catalyzes the conjugation of the bioactive-thiol glutathione (GSH) with a wide range of electrophilic substrates (nucleophilic displacement reactions). GST's are dimeric proteins with subunits of 25 to 29 kDa and can be found as multiple forms (isozymes), with various substrate specificity. GST-mediated conjugation is a major mechanism of detoxification in animals (*6*), plants (*7*), and certain microorganisms (*8,9*). Glutathione conjugation, catalyzed by the enzyme GST, has been shown to be a mechanism of plant metabolism and detoxification of various herbicides (*7*). These include chloroacetamides (acetochlor, alachlor, metolachlor, propachlor), diphenyl-ethers (acifluorifen, fluorodifen), triazines (atrazine, metribuzin), and thiocarbamates (EPTC, molinate).

Pioneering work by Lamoureaux (*10*) implicated the role of GST-mediated GSH conjugation in the metabolism of the herbicide propachlor in soil as well as plants. The role of GST-conjugation in the dechlorination of another chloroacetamide herbicide, acetochlor, in soil has also been established (*11*). Microbial products of GSH conjugates of the chloroacetamides in soil include the sulphonic, oxanillic and sulphinylacetic acid derivatives (*10,11*).

Most research characterizing the role of GST in bacterial detoxification has focused on non-herbicide substrates such as 1-chloro-2,4-dinitrobenzene [CDNB] (*8,12*). The role of GST in the dechlorination of dichloromethane (*13,14*) by a strain of *Methylobacterium* as well as a unique GSH-mediated reductive dechlorination of tetrachloro-*p*-hydroquinone by a strain of *Flavobacterium* (*15*) has recently been demonstrated. Unpublished research from our laboratories indicate that CDNB and alachlor-GST activity is observed in cell-free enzyme preparations from certain Gram-negative rhizosphere-competent bacteria. This paper addresses the potential of herbicides as substrates for rhizobacterial GST. These substrates include three groups of herbicides: chloroacetamides (alachlor), triazines (atrazine), and diphenyl-ethers (fluoridifen). We have also examined GST activity in bacterial rhizosphere communities with CDNB and alachlor as substrates.

Alachlor Metabolism by Rhizobacteria

Alachlor-GST activity has been demonstrated by various *Enterobacteriaceae* and *Pseudomonadaceae* strains in cell-free extracts (unpublished data). Time course studies of alachlor metabolism by cell suspensions were conducted to further elucidate the contribution of the GST pathway to the dechlorination of alachlor by these strains.

Cell suspensions (approximately log 10.5 cells ml^{-1}) were obtained from stationary tryptic soy broth (TSB) cultures of the following strains: *P. fluorescens* (UA5-40, BD4-13, and PRA25), *P. cepacia* (AMMD), *P. putida* (M-17), *Enterobacter cloacae* (ECH1), *Citrobacter diversus* (JM-92), *Klebsiella planticola* (JM-676) and *Serratia plymutica* (SP). Alachlor was added to 1.0 ml of cell suspension in KPi (potassium phosphate buffer, 0.1 M, pH 6.8) to achieve a final concentration of 100 μM (0.56 μCi, ^{14}C-ring labelled alachlor per tube) and were incubated for 96 h at 24°C on a rotary shaker. Samples (100μl) were removed at least every 24 h. The reaction was terminated by addition of 100 μl acetonitrile, which lyses the cells and aids in extraction of alachlor and metabolites. The supernatant fluid was recovered by centrifugation at 10,000 x g for 15 min.

The acetonitrile extract was spotted onto silica gel plates and developed with butanone (methyl ethyl ketone):acetic acid: water (10:1:1; solvent system A) for 10 cm. The plates were redeveloped for 5 cm from the original solvent front in a 180° rotation with hexane:methylene chloride:ethyl acetate (12:2:6: solvent system B). The distribution of radioactivity on the TLC plates was determined by linear scanning (Bioscan 200, Bioscan Inc, Washington, D.C.). Standards of alachlor-glutathione (AL-SG), alachlor-cysteineglycine (AL-CysGly), alachlor-cysteine (AL-Cys) and alachlor-sulfonic acid (Al-sulf) were synthesized according to methods described in the literature (16). All standards were at least 95% radiochemical purity. Typical Rf values following the second development were: AL-SG (0.03), AL-CysGly (0.10), AL-Cys (0.18), AL-sulf (0.40), alachlor (0.67). Alachlor initially developed at an Rf of 0.95 in solvent system A, however it migrated approximately 3 cm following the second 5 cm development with solvent system B. Verification of product identity in a third solvent system, C (butanol:acetic acid:water, 12:3:5;) and by HPLC/radioactivity detection was conducted for several samples. Data presented are typical observations for an experiment of 2 replicates. These experiments were repeated at least twice.

During a 96 h incubation, 22 to 100% of the alachlor initially present was metabolized by all strains tested except *P. cepacia* strain AMMD and *S. plymutica* strain SP. Representative patterns of alachlor metabolism for several strains are presented in Figure 1a-d. After the 96 h incubation, 75 to 100% of the alachlor was metabolized by active pseudomonads (Figure 1a -c), while only 22 to 35% of the alachlor was metabolized by the enterics: JM-676, JM-92 and ECH1 (Figure 1d). It should be noted that one *P. fluorescens* strain, UA5-40, dechlorinated 100% of the alachlor within 48 h (Figure 1a).

All alachlor-metabolizing bacteria produced two of the AL-SG metabolites, AL-CysGly and AL-Cys. AL-SG was occasionaly observed only in strain UA5-40. AL-CysGly was a transient minor metabolite in all strains except UA5-40, in which it accounted for 60% of the alachlor at 24h. In the other strains, AL-Cys was the major accumulating intermediate which may be due to higher alachlor-GST activity and lower alachlor-GGTP (GGTP = γ-glutamyltranspeptidase) activity in UA5-40 than in the other pseudomonads tested.

All alachlor metabolizing strains produced an unidentified moderately polar compound (Rf of 0.95, solvent system A) as the major end-product. This metabolite was immobile in solvent system B. Two *P. fluorescens* strains (UA5-40 and PRA25) accumulated a second final product with an Rf of approximatley 0.75 in solvent system A. This product was also immobile in solvent B. We are presently attempting to identify both compounds. A series of potential products can be derived from GSH-conjugation with alachlor as presented in the metabolic pathway summarized in Figure 2, 3 and 4.

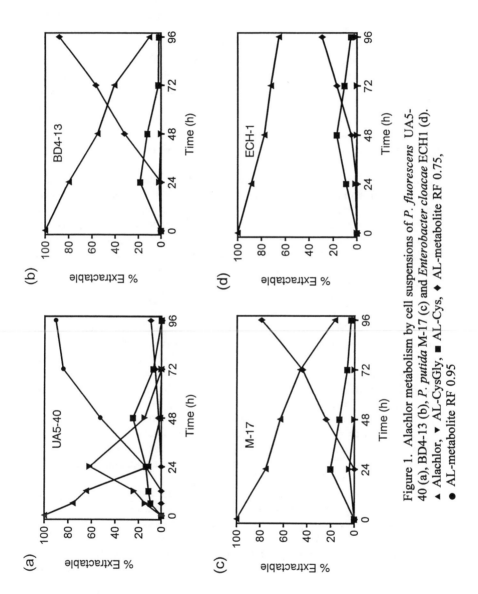

Figure 1. Alachlor metabolism by cell suspensions of *P. fluorescens* UA5-40 (a), BD4-13 (b), *P. putida* M-17 (c) and *Enterobacter cloacae* ECH1 (d). ▲ Alachlor, ▼ AL-CysGly, ■ AL-Cys, ♦ AL-metabolite RF 0.75, ● AL-metabolite RF 0.95

The growth of 12 strains of rhizosphere competent bacteria was assessed in TSB amended with 0, 50, 100, 200 and 400 μM alachlor. The highest concentration (400 μM) represents the maximum solubility of alachlor in water. All alachlor metabolizing strains tested (UA5-40, PRA-25, M-17, BD4-13, ECH1, JM-92) exhibited similar growth patterns (generation times and final cell yield) at all concentrations of alachlor as in unamended broth . All four strains incapable of metabolizing alachlor (*P. cepacia* - AMMD, *S. plymutica* - SP, and two bacilli, (*B. subtilis* -GBO7 and *B. thuringiensis*- HD-2) exhibited a 20 to 50% reduction in generation time during the logarithmic growth stage at alachlor concentrations greater than 50 or 100 μM compared to unamended broth. However only *B. subtilis* strain GB-07 exhibited a 50% reduction in final yield at all alachlor concentrations compared to unamended broth.

Hydrolytic Enzymes and Glutathione Conjugate Metabolism.

In most organisms, glutathione conjugates are further catabolized by sequential hydrolysis of amino acids from the GSH tripeptide (Figure 2). The first peptidase, γ-glutamyltranspeptidase, is specific for cleavage of the glutamic acid moiety, followed by a second carboxypeptidase- or peptidases- cleaving glycine from the CysGly-conjugate. Several possibile pathways for further catabolism of the Cys-conjugate exist. A family of cysteine-β-lyases (*17, 18,19 20, 21*) can vary in specificity with the products being the corresponding mercaptan, ammonia, and pyruvate (Figure 3). In mammalian systems the activities of GST and the downstream catabolic enzymes are compartmental in various organs (*11, 18*), while bacterial enzymes are observed in the cytosol.

Alachlor-GGTP activity and subsequent metabolism of AL-CysGly in cell-free enzyme preparations of various rhizobacteria was studied. Assay mixtures included cell-free extract (3 to 4 mg protein), 200 nmoles AL-SG (0.06 μCi ^{14}C-ring labeled AL-SG), and 100 μmole KPi (pH 6.8) in a final volume of 1.0 mL. Incubations were conducted for 4 h at 25°C, with 100 μl samples removed at 1, 2 and 4 h. The assay was terminated by adding 100 μl acetonitrile, and processed as described in the whole cell-alachlor metabolism study.

The ability to degrade the AL-SG conjugate was found in all rhizobacterial strains evaluated, regardless of their ability to metabolize alachlor (Table I). *P. putida* (M-17) and *P. fluorescens* strain BD4-13 completely degraded the AL-SG conjugate to AL-CysGly in 1 h with 40 to 50% of the AL-CysGly degraded to AL-Cys within 4 h. Activity of the carboxypeptidase, responsible for glycine hydrolysis, was detected in all strains tested, but was ten-fold slower than AL-SG-GGTP activity.

Cysteine-β-lyase activity was determined in cell-free extracts of rhizobacterial strains using S-benzyl-L-cysteine, cystathionine or S-ethyl-L-cysteine as substrate. The assay was based upon production of pyruvate (*20, 21*)

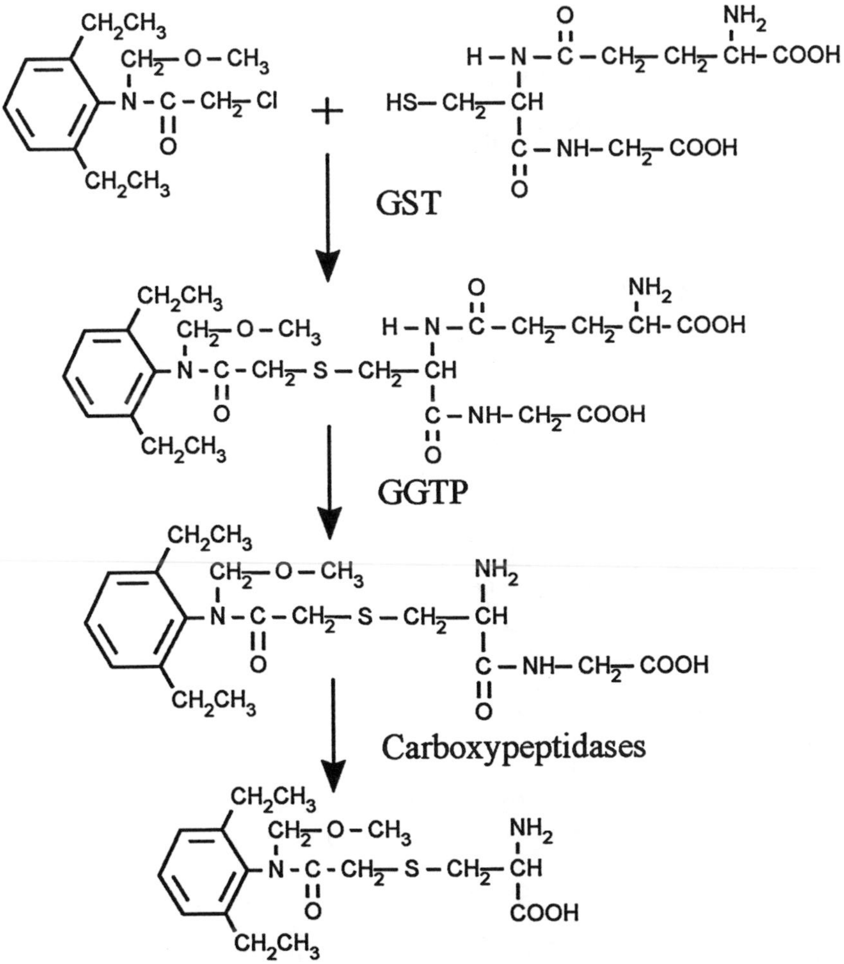

Figure 2. Catabolic pathways for glutathione conjugation of alachlor and initial catabolism of the glutathione conjugate.

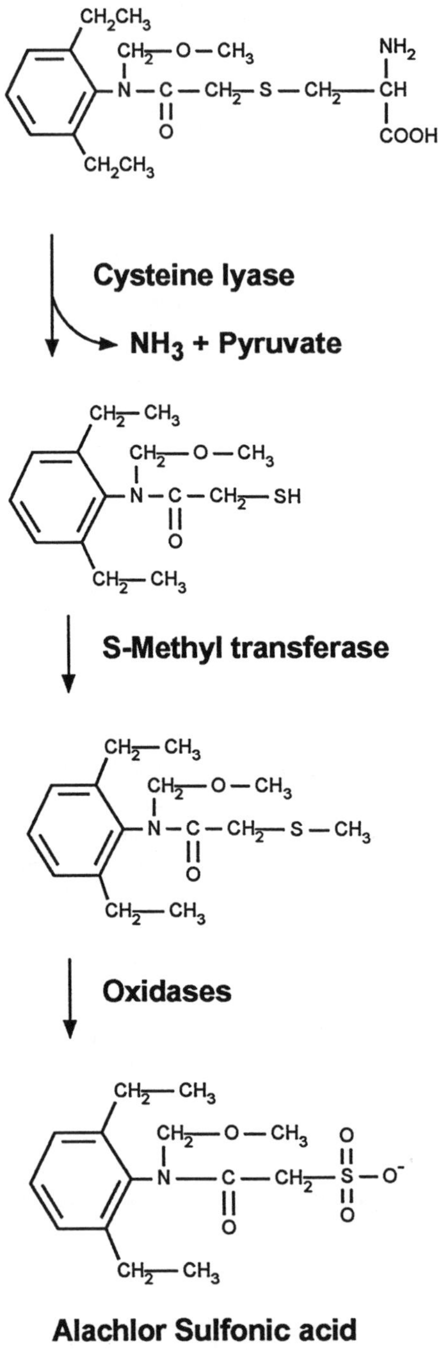

Figure 3. Catabolic pathway of alachlor-cysteine based upon cysteine β-lyase.

Table I. Alachlor-γ-glutamyltranspeptidase Activity and Subsequent Al-CysGly Catabolism by Cell-free Extracts of Rhizobacterial Strains

Bacterial Strain	AL-SG GGTP activity[1] nmoles mg^{-1} protein h^{-1}	AL-Cys formed nmoles after 4 h
M-17	>50	105 ± 8[2]
BD4-13	47 ± 5	82 ± 7
UA5-40	32 ± 4	10 ± 3
PRA25	41 ± 3	31 ± 6
ECH1	28 ± 6	11 ± 4
AMMD	38 ± 4	5 ± 5
JM92	24 ± 5	12 ± 5

[1]Activity determined following 1 h incubation.
[2]Mean and standard deviation of three replicates.

and the reaction mixture consisted of substrate (1 μmole), NADH (0.5 μmole), pyridoxal phosphate (0.125 μmole), lactate dehydrogenase (10 units), KPi (100 μmole, pH 8.0), and 3 to 4 mg of protein (cell-free extract) in a final volume of 1.0 ml. The reaction mixture (minus substrate) was incubated 2 min prior to adding substrate. The decrease in absorbance at 340 nm (NAD formation) was measured spectrophometrically for 90 sec. Control reactions (all components except substrate) were determined for each strain and included in the corrected calculation of specific activity.

Cysteine-β-lyase activity was detected in all strains tested (Table II). However, the activity was substrate-dependent, i.e., some strains had greater alkyl versus aryl cysteine-β-lyase activity. Strain UA5-40 had the greatest activity on all three substrates, while *P. cepacia*, AMMD and *E. cloacae*, ECH1 had only a trace of activity.

Activities of cysteine-β-lyase observed in these studies were similar to those observed for certain intestinal bacteria and enrichments from aquatic and terrestrial environments (*17*). Cystathione-β-lyase is a component of the methionine biosynthesis pathway and would be expected to be present in most bacteria. The products of cysteine-β-lyase, mercaptans, are short lived and readily methylated and / or oxidized (*17*). Identification of end- products of alachlor metabolism can aid in elucidating the role of cysteine-β-lyases in metabolism of herbicide-SG conjugates in rhizobacteria. The β-cleavage of the C-S bond is only one of several potential metabolic fates of the cysteine conjugate. Other possible pathways of the cysteine conjugate are outlined in Figure 4. An α-cleavage of the C-S bond can give rise to the corresponding acetamide or hydroxyacetamide. The hydroxyacetamide can be further oxidized to the oxanilic acid. Deamination of the cysteine conjugate results in the sulfinyl acetic acid.

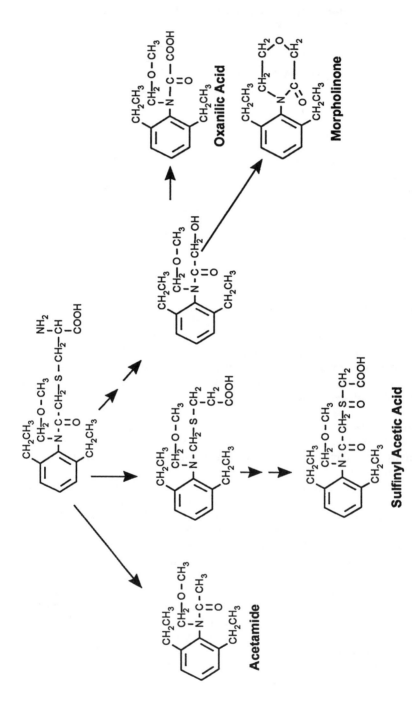

Figure 4.   Alternative catabolic pathways of alachlor-cysteine.

Table II. Cysteine-β-lyase Activity of Cell-free Extracts of Rhizobacterial strains

Bacterial	nmoles mg⁻¹ protein min⁻¹		
Strain	Ethyl-Cys	Benzyl-Cys	Cystathionine
UA5-40	1.6 ± 0.5^1	1.6 ± 0.4	0.8 ± 0.3
PRA25	1.3 ± 0.6	0.2 ± 0.1	0.6 ± 0.1
BD4-13	1.0 ± 0.2	0.4 ± 0.2	0.6 ± 0.2
M-17	0.8 ± 0.3	0.4 ± 0.1	0.4 ± 0.1
AMMD	< 0.1	< 0.1	0.2 ± 0.1
ECH1	< 0.1	< 0.1	0.2 ± 0.1

[1]Mean and standard deviation of 3 or 4 replicates

CDNB-GST and Alachlor Metabolism of Rhizobacterial Populations

The CDNB-GST activity and potential for alachlor metabolism was studied in enrichment cultures of rhizosphere populations, similar to procedures used to study cysteine-β-lyase in environmental populations (*17*). One gram of rice (greenhouse-grown) or soybean (field-grown) roots were shaken in 100 ml of 0.1 M KPi (pH 7.0) for 30 min and 5.0 ml was transferred to 400 ml of dilute (10%) TSB. The cultures were incubated for 72 h on a shaking incubator (24° C, 150 rpm). Cells were harvested by centrifugation, rinsed twice in KPi (0.1 M, pH 6.8) and brought to a final cell density of log 10.5 ± 0.3 cells ml⁻¹. Cell counts were determined by serial dilution and plating on 10% tryptic soy agar (TSA) or crystal violet (2.0 μg ml⁻¹) TSA (total and Gram-negative counts respectively). CDNB-GST activity was determined on cell-free extracts (*8*). Alachlor metabolism by whole cells was determined after a 24 h incubation as described previously.

A range of CDNB-GST activity comparable to UA5-40 and M-17 was observed for most rhizosphere enrichments (Table III). Alachlor metabolism via the GST-pathway was also observed in all enrichments. In the rice samples, AL-Cys was the major metabolite (>75%), AL-CysGly was detected in all four rice rhizosphere enrichments, but AL-SG was not. In the soybean cultures alachlor-SG was the major product observed (41 to 76%) with the AL-Cys (10 to 36%) conjugate observed in all cultures. All rice and soybean rhizosphere cultures exhibited a lower level of alachlor metabolism compared to the more active alachlor catabolizing *Pseudomonas* strains. Gram-negative bacteria accounted for 44 to 77% of these enriched rhizosphere populations. In most studies Gram-negative bacteria are several fold higher compared to root-free soil (*22*, Hoagland et al., this volume).

Formation of AL-SG and / or the subsequent metabolic products AL-CysGly or AL-CYS in rhizosphere cultures from both rice and soybeans indicate that GSH conjugation may have ecological significance in the detoxification/dechlorination of alachlor in rhizobacterial communities.

Individual bacterial strains capable of mineralizing alachlor have not been reported (*23, 24*), however these studies have been successful in the isolation of bacteria capable of degrading alachlor to polar metabolites. GSH-mediated catabolic pathways are one of several possible pathways for the biodegradation of chloroacetamide herbicides. Other possibile pathways include oxidation of alkyl chains (*25, 26*) and arylacylamidase-mediated hydrolysis (*24*). Arylacylamidase-mediated hydrolysis of propachlor by mixed bacterial communities has been reported (*24*). A unique amidase cleavage of propachlor by a species of *Moraxella* forming 2-chloro-N-isopropylacetanilide and a catechol has also been reported (*27*).

Table III. CDNB-GST Activity of Cell-free Extracts and Whole Cell Alachlor-GST Activity of Rhizosphere Enrichments from Rice and Soybean Roots, and Reference Rhizobacteria Cultures

Culture	CDNB-GST $nmol\ mg^{-1}\ protein\ h^{-1}$	Alachlor-GST $nmol\ mg^{-1}\ protein\ 24h^{-1}$
Rice A (75% GNB)[1]	1.2 ± 0.5 [2]	1.2 [3]
Rice B (47% GNB)	1.4 ± 0.5	1.4
Rice C (77% GNB)	1.2 ± 0.1	1.0
Rice D (63% GNB)	0.8 ± 0.2	0.6
Soybean A (67% GNB)	0.8 ± 0.2	1.3
Soybean B (39% GNB)	0.5 ± 0.1	1.1
Soybean C (53% GNB)	0.7 ± 0.1	0.6
Soybean D (49% GNB)	1.2 ± 0.2	0.8
UA5-40	0.9 ± 0.3	6.7
PRA-25	2.8 ± 0.4	2.1
M-17	1.4 ± 0.4	2.4
ECH1	2.9 ± 0.5	1.5

[1]%GNB = % Gram-negative bacteria of the total bacterial population.
[2]Mean and standard deviation of 3 replicates.
[3]Average of two replicates based upon total AL-SG, AL-CysGly and AL-Cys formed during a 24 h incubation.

Atrazine-GST Studies

Atrazine-GST assays were conducted on cell-free extracts using a phase-partitioning assay modified from Lamoureaux and Rusness (*28*). Assay reaction mixtures included 50 nmoles atrazine (25 nCi, ^{14}C-ring-labelled atrazine diluted with nonlabelled atrazine), GSH 2.0 μmoles, 2 to 10 mg protein (cell-free extract), KPi 100 μmoles, pH 6.8 in a final volume of 600 μl. Following incubations of 30 min to 4 h reactions were terminated by the addition of 600μl of methylene chloride. Phase separation was enhanced by centrifugation at 14,000 x g for 5 min and radioactivity in the aqueous fraction was determined by liquid scintillation counting. Controls minus enzyme preparations were also included. Aliquots were spotted on silica gel TLC plates, developed with solvent system C, and analyzed for distribution of radioactivity by linear scanning. Strains tested include *Pseudomonas* strains (UA5-40, BD4-13, M-17, AMMD and PRA25), *E. cloacae* strains (EC399 and ECH1), and strains of *Alcaligines, Arthrobacter, Citrobacter, Klebsiella, Serratia and Xanthomonas*. Atrazine metabolism by whole cells was determined by addition of 50 nmoles atrazine (0.4 μCi diluted with nonlabelled atrazine) to 1.0 ml of cell suspensions of the above strains containing approximately log 10.5 cells ml[-1] as in the previously described alachlor metabolism studies.

   Atrazine-GST activity was not observed in any bacterial strains tested to date in our studies with whole cells or cell-free extracts. We did not observe metabolites of atrazine (based upon the TLC analysis) by whole cells of any strain during a 72 to 96 h incubation. In most studies of atrazine degradation by various soil bacteria, dealkylation appears to be the major degradation pathway (*29, 30, 31*). Dealkylation may likewise precede dechlorination of atrazine, although the dechlorinated product hydroxy-atrazine is found in soils (*31*). A recent report (*32*) has indicated that rapid mineralization of atrazine can be observed by mixed cultures from enrichment procedures, the dechlorinated product hydroxyatrazine is one of the intermediates observed in this study.

Fluorodifen-GST activity

Cell-free extracts were prepared as described in the previous studies except that pH 8.0 KPi was used. Enzyme assays were modified from Frear and Swanson (*33*) and Lamoreaux and Rusness (*28*). The assay mixture included cell-free extract (3 to 5 mg protein), KPi 100 μmoles, pH 8.0, fluorodifen 100 nmoles (added in 25μl of acetone) and 2.0 μmoles glutathione. Changes in absorbance at 405 nm were observed over a 30 min incubation at 25 C. Production of the hydrolysis product *p*-nitrophenol was verified by TLC. Due to unavailability of labelled material, verification of the GSH conjugate was not feasible. The assay mixture of three to four replicates were combined and extracted twice with 10 ml of ethyl ether. The ether extracts were combined, evaporated to dryness under N_2 and dissolved in 25 μl of methanol. The methanol extract was spotted on silica gel plates and developed for 10 cm with benzene:acetic acid (100:4, solvent system D).

Fluorodifen-GST activity, based upon an increase in absorbance at 405 nm (p-nitrophenol production), was observed for the limited strains tested (Table IV). Formation of the hydrolysis product, p-nitrophenol, during a 30 min incubation, was verified for three of the more active strains (M-17, AMMD and PRA25) by TLC.

Table IV. Fluorodifen-GST Activity in Cell-free Extracts of Rhizosphere-Competent Bacteria

Strain	nmoles mg^{-1} protein h^{-1}
Pseudomonas	
M-17	1.86 ± 0.02[1]
AMMD	1.12 ± 0.06
PRA25	0.60 ± 0.09
BD4-13	0.19 ± 0.02
UA5-40	0.18 ± 0.02
Enterobacter	
EC39979	0.22 ± 0.04

[1] Mean and standard deviation of four replicates.

Conclusions and Relevance to Bioremediation

The GST-pathway of metabolism / detoxification by rhizobacteria has been demonstrated for two families of herbicides (chloroacetamides and diphenyl ethers) and the non-herbicide substrate CDNB. We have also demonstrated GST-activity in bacterial rhizosphere communities and the role of GSH-conjugation in the degradation/detoxification of alachlor. Rates of GST-mediated dechlorination of alachlor by reference strains and rhizosphere communities were several orders of magnitude lower than optimized *in vitro* rates reported in mammalian liver cytosol fractions (0.6 to 6 nmoles mg^{-1} protein 24 h^{-1} versus 11 to 70 nmoles mg^{-1} protein min^{-1} , respectively [16]). The abundance of Gram-negative bacteria in plant rhizospheres relative to soil (22) can favor the rhizosphere as a site for bacterial GST-mediated detoxification or herbicides and other xenobiotics.

Numerous bacterial strains have been isolated for various bioremediation strategies (34, 35). Poor ecological fitness upon reintroduction of the degrading bacteria to the environment has been one limiting factor in the use of certain bacterial inoculants for bioremediation (35). Dicamba-degrading bacteria have been introduced by soil inoculation to protect susceptible species from herbicide toxicity (*36*). Detoxification has been proposed as a potential mode of action of plant growth-promoting rhizobacteria (*37*). The alachlor-degrading *Pseudomonas* spp. described here have been characterized as either plant growth-promoting rhizobacteria or biological control agents [M-17 (*38*), PRA25 (*39*), BD4-13 (*40*), and UA5-40 (Eugene Milus, personal communication])], with excellent rhizosphere competence. Fluorescent pseudomonads can be introduced as seed inoculants and be dispersed through the soil by both root colonization and water percolation. The strains described here have high tolerance for alachlor (400 μM) and moderate to good detoxification potential (several-fold higher than indigenous rhizosphere bacteria) in addition to rhizosphere competence. These strains have been identified from a survey to ascertain the potential for GST-mediated detoxification. Bacterial isolates with greater potential for degradation of alachlor and other GST-substrates, may be derived from a greater selection pressure.

Acknowledgments

We appreciate the excellent technical contributions of R.E Gordon and M.E. Smyly. We are extremely grateful to the following individuals for contributing bacterial strains used in these studies: Charles Hagedorn - BD4-13 (V.P.I.S.U.), Charles Howell - ECH1 (USDA-ARS, College Station), Mark Mount - M-17 (Univ Mass, Amherst), John MacInroy and Joe Kloepper - (Many eneterics JM-92 and JM676 (Auburn Univ.), and Jennifer Parke - PRA25 and AMMD (Univ. Wisconsin) and Stan Nemec - SP (USDA-ARS, Orlando). We also wish to thank Louis Gaston, Krishna Reddy, Joe Lepo, and Charles Hagedorn for their critical reviews of this manuscript.

Literature Cited

1. Rovira, A.D.; Davey, C.B. In *The Plant Root and its Environment*. Carson, E.W., Ed., Univ. Press: Charlottesville, VA, **1974**, pp. 153-204.
2. Rovira, A.D.; *The Soil-Root Interface*. J.L. Harley, R.S. Russell ed. Academic Press, NY, **1979**, pp. 145-160.
3. Hsu, T.S.; Bartha, R. *Appl. Environ. Microbiol.* **1979**, *36*, 36- 41.
4. Reddy, B.R.; Sethunathan, N. *Appl. Environ. Microbiol.* **1983**, *45*, 826-829.
5. Walton, B.T.; Anderson T.A. *Appl. Environ. Microbiol.* **1990**, *56*, 1012-1016.
6. Habig, W.H.; Pabst, M.J.; Jakoby, W.B. *J. Biol. Chem.* **1974**, *249*, 7130-7139.
7. Lamoureaux, G.L.; Rusness, D.G. In *Glutathione: Chemical, Biological and Medical Aspects; Dolphin, D.; Poulson, R.; Avramovic, O, Ed.;* John Wiley & Sons: New York, New York, **1989**, Part B; pp. 154-196.
8. Lau, E.P.; Niswander, L.; Watson, D.; Fall, R.R. *Chemosphere.* **1980**, *9*, 733-740.

9. Cohen, E.; Gamliel, A.; Katan, J. *Pestic. Biochem. Physiol.* **1986,** *26,* 1-9.
10. Lamoureaux, G.L.; Rusness, D.G. *Pestic. Biochem. Physiol.* **1989,** *34,* 187-204.
11. Feng, P.C.C. *Pestic. Biochem. Physiol.* **1991,** *40,* 136-142.
12. Shishido, S. *Agric. Biol. Chem.* **1981,** *12,* 2951-2953.
13. LaRoche, S.D.; Leisinger, T. *J. Bacteriol.* **1990,** *172,* 164-171.
14. Leisinger, T.; Kohler-Staub, D. *Methods Enzymol.* **1990,** *188,* 355-361.
15. Xun, L.; Topp, E.; Orser, C.S. *J. Bacteriol.* **1992,** *174,* 8003-8007.
16. Feng, P.C.C.; Pantella, J.E. *Pestic. Biochem. Physiol.* **1988,** *31,* 84-90.
17. Larsen, G.L. *Xenobiotica.* **1985,** *15,* 199-209.
18. Larsen, G.L. *Xenobiotica.* **1981,** *11,* 473-480.
19. Nomura, J.; Nishizuka, Y.; Hayaishi, O. *J. Biol. Chem.* **1963,** *238,* 1441.
20. Guggenheim, S. *Meth. Enzymol.* **1971,** *17B,* 439-442.
21. Larsen, G.L.; Stevens, J.L. *Mol. Pharmacol.* **1986,** *29,* 97-103.
22. Holding, A.J. *J. Appl. Bact.* **1960,** *3,* 515-525.
23. Felsot, A.S.; Dzantor, E.K. In *Enhanced Biodegradation of Pesticides;* Racke, K.D. Coats, J.R. Ed.; ACS Books, Wash. D.C. **1990,** pp. 294-268.
24. Novick, N.J.; Alexander, M. *Appl. Environ. Microbiol.* **1985,** *49,* 737-743.
25. Feng, P.C.C.; Patanella J.E. *Pestic. Biochem. Physiol.* **1989,** *33,* 16-25.
26. Pothulari, J.V.; Freeman, J.P., Evans, F.E.; Moorman, T.B.; Cerniglia, C.E. *J. Agric. Food Chem.* **1993,** *41,* 483-486.
27. Villareal, D.T.; Turco, R.F.; Konopoka, A. *Appl. Environ. Microbiol.* **1991,** 2135-2140.
28. Lamoureaux, G.L.; Rusness, D.G. *Pestic. Biochem. Physiol.* **1986,** *26,* 323-342.
29. Erickson, L.E.; Lee, K.H. *Crit. Rev. Environ. Control.* **1989,** *19,* 1-14.
30. Behki, R.M.; and S.U. Khan. *J. Agric. Food Chem.* **1986,** *34,* 746-749.
31. Behki, R.; Topp, E.; Dick, W.; Germon, P. *Appl. Environ. Microbiol.* **1993,** *59,* 1955-1959.
32. Mandelbaum, R.T.; L.P. Wackett; D.L. Allan. *Appl. Environ. Microbiol.* **1993,** *59,* 1695-1701.
33. Frear, D.S.; Swanson, H.R. *Pestic. Biochem. Physiol.* **1983,** *3,* 473-482.
34. Boyle, M. *J. Environ. Qual.* **1989,** *18,* 395-402.
35. MacRae, I.C. *Rev. Environ. Contam. Toxicol.* Ware, G.W. ed. Springer-Verlag: New York, New York, **1989,** Vol. 109; pp. 1-87.
36. Krueger, J.P.; Butz, R.G., and Cork, D.C. *J. Agric Food Chem.* **1991,** *39,* 1000-10003.
37. Kloepper, J.W.; Lifshitz, R.L.; Zablotowicz, R.M. *Trends Biotechnol.* **1989,** *7,* 39-44.
38. Grimes, H.D.; Mount, M.S. *Soil Biol. Biochem.* **1984,** *16,* 27-30.
39. Parke, J.L.; Rand, R.E.; Joy, A.E.; and King, E.B. *Plant Disease.* **1991,** *75,* 987-992.
40. Hagedorn, C.; Gould, W.D.; Bardinelli, T.R. *Plant Disease.* **1993,** *77,* 278-282.

RECEIVED February 28, 1994

Chapter 16

Biological Degradation of Pesticide Wastes in the Root Zone of Soils Collected at an Agrochemical Dealership

Todd A. Anderson, Ellen L. Kruger, and Joel R. Coats

Pesticide Toxicology Laboratory, Department of Entomology, Iowa State University, Ames, IA 50011–3140

Evidence for enhanced microbial degradation of xenobiotic chemicals in the rhizosphere, a zone of increased microbial activity at the root-soil interface, continues to accrue, suggesting that vegetation may play an important role in facilitating bioremediation of contaminated surface soils. For sites tainted with pesticide wastes, such as at agrochemical dealerships, establishing vegetation may be problematic because of the presence of herbicide mixtures at concentrations severalfold above field application rates. Nonetheless, herbicide-tolerant plants exist that can survive in these environments, and they are ideal candidates for testing the influence of rhizosphere microbial communities on the degradation of pesticide wastes. Experiments in this laboratory have tested whether a commodity plant such as soybean could survive in soil from a pesticide-contaminated site containing a mixture of three predominant herbicides, atrazine, metolachlor, and trifluralin, and if its presence could enhance biodegradation. Although soybean survival in this soil was high, its presence did not enhance the degradation of the chemicals. Tests with nonvegetated soils and rhizosphere soils from *Kochia* sp., a herbicide-tolerant plant, showed enhanced degradation of these chemicals in rhizosphere soil. Also, *Kochia* sp. seedlings have emerged from rhizosphere soils spiked with additional concentrations of the three test chemicals, indicating the ability of these plants to survive in soils containing high concentrations of herbicide mixtures.

The plant rhizosphere (Figure 1) sustains microbial populations an order of magnitude or more above populations in unvegetated soil by secreting mucilaginous substances collectively known as root exudate (*1*). In addition, the rhizosphere fosters interactions among populations of microorganisms at the molecular, physiological, and ecological levels. These observations suggest that the dense, diverse, and synergistic microbial community in the rhizosphere may contribute to greater rates of metabolism of xenobiotic compounds compared with microbial metabolism in unvegetated soils. Seminal work on the fate of pesticides in the rhizosphere by Hsu and Bartha (*2*), Reddy and Sethunathan (*3*), Sandmann and Loos (*4*), and Lappin and coworkers (*5*)

0097–6156/94/0563–0199$08.00/0

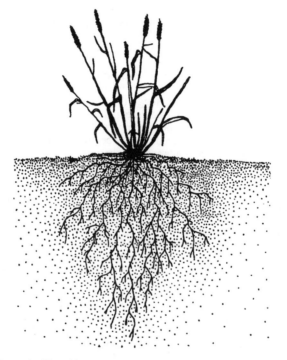

Figure 1. The rhizosphere, or root-soil interface, of a grass illustrating the gross morphology of a fibrous root system.

provided precedent for enhanced microbial degradation in the root zone and subsequently was further supported in similar studies with industrial chemicals by Rasolomanana and Balandreau (6), Aprill and Sims (7), Walton and Anderson (8), and Knaebel and Vestal (9). The previous research on enhanced microbial degradation of organic compounds in the rhizosphere has been recently reviewed (10, 11,12). In this paper, we review the literature germane to selecting appropriate plant species for enhancing microbial degradation at sites contaminated with pesticide wastes. In addition, our present research on microbial degradation of pesticides in rhizosphere and nonvegetated soils from a herbicide-contaminated site in Iowa is described.

Microorganism-Plant-Chemical Interactions in Soil

Critical to the concept of using vegetation to enhance microbial degradation is the ability of plants to survive in the contaminated environment. For sites tainted with industrial chemicals such as chlorinated solvents or petroleum hydrocarbons, the toxic effect of the chemical on the plant usually does not prohibit establishment. However, in pesticide-contaminated sites, in which mixtures of herbicides can be present at concentrations severalfold above field application rates, establishing vegetation may be more problematic. Not only are there high concentrations of chemical mixtures, but also the chemicals present are designed to inhibit the growth of vegetation. Nonetheless, herbicide-tolerant plants that can survive in these soils exist. Regardless of the type of chemical contamination, plants present at these contaminated sites are ideal candidates for testing the influence of the rhizosphere on microbial degradation of hazardous organic compounds. Clearly, plants that can survive in these environments have been selected and have special mechanisms for dealing with the potentially detrimental effects of the chemicals.

The interaction between plants and microbial communities in the rhizosphere is a complex association that has evolved to the mutual benefit of both groups. In addition to the established relationships between plants and soil microorganisms regarding crop productivity, other relationships undoubtedly exist, such as the potential role of the rhizosphere microbial community in protecting the plant from chemical injury (*Walton et al., this volume*). Previous research has shown that plants increase root exudation in the presence of xenobiotic chemicals (13, 14). In hydroponic cultures of corn, the presence of simazine (2-chloro-4,6-bis-ethylamino-s-triazine), a preemergence herbicide used for controlling weeds in corn, caused a twofold increase in exudation of organic acids (15). In addition, simazine increased the length and weight of roots, but only if microorganisms also were present in the medium. It is not clear whether the increase in exudation is an evolved response by the plant to attract and nourish more microorganisms (and possibly accelerate degradation) or simply a physiological effect of the chemical on the plant. The tolerance of corn to the herbicidal effects of simazine may be predominantly the result of rapid metabolism of the compound by the plant. Most herbicides used to control weeds are readily metabolized by nontarget plants. Nonetheless, rhizosphere microbial communities may also play a role in protecting the plant from chemical injury. This idea is further supported by the works of Herring and Bering (16) and Krueger et al. (17). Herring and Berring (16) found that the toxic effect of phthalate esters on spinach and pea seedlings could be abated or reversed by the presence of microorganisms in the soil. In a similar study, Krueger and coworkers (17) showed that three strains of microorganisms capable of degrading dicamba (3,6-dichloro-2-methoxybenzoic acid) could be used to reduce the herbicidal activity of dicamba in the rhizosphere of peas at 8-fold above the normal field application rate. This rapid removal of dicamba was such that this susceptible plant could then grow in the treated soil.

The ability of the plants to select for different microbial communities in both composition and size is intriguing from the standpoint of exploring whether this selection translates into differences in the rates of microbial degradation of organic compounds in the rhizosphere. The literature suggests that this is the case for both industrial and agrochemicals; however, the *raison d'être* for these selections seems to differ. For sites contaminated with herbicides, plants that can enhance the size or stature of the degradative microbial community in the root zone seem less sensitive to the chemical. Indeed, the degradation of certain herbicides is enhanced in the rhizosphere of plants that are relatively insensitive to the herbicidal effects of the chemical (Table 1). For example, Sandmann and Loos (4) observed an increase in the numbers of 2,4-D-degrading microorganisms in sugarcane rhizosphere compared with African clover. The authors suggested that the increase in degraders was a possible mechanism by which sugarcane is protected from the toxic effects of 2,4-D and that phenolic analogs in the exudate selected for the microbial community responsible for degrading 2,4-D. The interaction of leguminous plants with nitrogen-fixing bacteria results in increased microbial biomass, plant growth, and root exudation, perhaps because of the increased availability of soil nitrogen in the presence of nitrogen-fixing bacteria. This, in turn, may lead to enhanced microbial degradation of herbicides in the rhizosphere by these bacteria (18). With insecticides, toxicity of the chemical to the plant usually is not an issue. Thus, the degradation of insecticides such as diazinon and parathion, which are cometabolized, is accelerated in the rhizosphere, probably because of the large microbial community present and its enhanced cometabolic activity rather than because of the composition of that community (2).

Plants may play a completely different role in sites contaminated with industrial chemicals. For example, legumes provide an important source of nitrogen for microbial degradation in petroleum-contaminated sites (19). In addition, the plant may select for a certain type of microbial community capable of degrading the contaminant(s), but selection is not as critical for plant survival as is the case with herbicide-contaminated sites. Anderson and Walton (20) observed differences in microbial degradation of trichloroethylene (TCE) in the rhizospheres of five plant species. In whole-plant studies, the mineralization of ^{14}C-TCE was enhanced in three of the species tested (*Pinus taeda*, *Lespedeza cuneata*, and *Glycine max*) although only two (*P. taeda* and *L. cuneata*) were indigenous to the contaminated site. Degradation rates for TCE in the rhizospheres of *Solidago* sp. and *Paspalum notatum*, both present at the site, were not significantly different from TCE-degradation rates in nonvegetated soil from the site. In addition, toxicity studies with TCE were conducted on soybean as well as rhizosphere soil samples from each of the plants. Extremely high TCE concentrations (400 µg/g) were needed to significantly reduce soybean net photosynthesis, whereas plant biomass production was unaffected in soil containing TCE as high as 1600 µg/g (ppm) (21). In most instances, microbial respiration in soil samples from the rhizosphere was unaffected by TCE concentrations as high as 500 µg/g.

Current Research

Background. Since the early 1950s, production and use of pesticides in agriculture has dramatically increased. Coupled with the increase in pesticide usage has been the rapid growth of retail agrochemical dealerships. Unfortunately, many of these dealerships have, through normal operating procedures, contaminated the soil and water at these sites. It is estimated that most dealerships throughout the Midwest have some type of problem from chemical contamination, creating one of the most ominous issues facing the agrochemical industry (22, 23, 24).

The expense of most of the current technologies for cleanup of contaminated soil and water preclude their use at agrochemical dealership sites. In addition, dealers

Table 1. Studies germane to selecting appropriate plant species for enhancing microbial degradation in the rhizosphere at pesticide contaminated sites

Plant	Family	Chemical	Comments	Reference
grasses	Gramineae	atrazine	Enrichment cultures from rhizosphere soils of grasses collected near the boundaries of a pesticide-contaminated site mineralized 60% of atrazine added at 100 ppm (0.46mM) after eight days, although a lag period of three days was observed.	(25)
wheat	Gramineae	mecoprop 2,4-D MCPA	Microbial community isolated from wheat roots and composed of five species was capable of using compounds as a carbon source. However, none of the pure cultures was capable of growth on mecoprop. Wheat is tolerant to this class of herbicides (phenoxy acids).	(5)
sugarcane African clover	Gramineae Fabaceae	2,4-D	Higher population of 2,4-D-degrading microorganisms in the rhizosphere of sugarcane, a plant tolerant to 2,4-D, compared with African clover, a plant sensitive to the herbicidal effects of 2,4-D.	(4)
bush bean	Fabaceae	diazinon parathion	Increased mineralization of both compounds in the rhizosphere of hydroponic bush bean. Because both compounds are insecticides, plant toxicity was not a problem. Both compounds are suspected of being cometabolized, a process that would be expected to occur very readily in the rhizosphere.	(2)
rice	Gramineae	benthiocarb	Eightfold increase in heterotrophic bacteria in the rhizosphere of treated rice plants.	(26)
flax	Linaceae	2,4-D	Ammonifying, nitrifying, and cellulose-decomposing bacteria in the rhizosphere increased by 1 to 2 orders of magnitude.	(27)

Continued on next page

Table 1. *Continued*

Plant	Family	Chemical	Comments	Reference
corn	Gramineae	atrazine	Increase in production of atrazine metabolites by rhizosphere microorganisms, especially in the presence of decomposing roots. Corn is relatively insensitive to the herbicidal effects of atrazine.	(28)
---	---	endosulfan DDT dieldrin PCNB PCP	A rhizosphere-competent fungus, *Trichoderma harzianum*, isolated from a banana plantation soil in Costa Rica degraded endosulfan, DDT, dieldrin, PCNB, and PCP *in vitro*.	(29)

2,4-D   2,4-dichlorophenoxyacetic acid
atrazine   2-chloro-4-ethylamino-6-isopropylamino-s-triazine
benthiocarb   S-p-chlorobenzyl diethylthiocarbamate
DDT   1,1,1-trichloro-2,2-bis(chlorophenyl)ethane
diazinon   O,O-diethyl-O-(2-isopropyl-6-methyl-4-pyrimidinyl) phosphorothioate
dieldrin   1,2,3,4,10,10-hexachloro-6,7-epoxy-1,4,4a,5,6,7,8,8a-octahydro-1,4-endo-exo-5,8-dimethanophthalene
MCPA   2-methyl-4-chlorophenoxyacetic acid
mecoprop   2-(2-methyl-4-chlorophenoxy)propionic acid
parathion   O,O-diethyl-O-p-nitrophenyl phosphorothioate
PCNB   pentachloronitrobenzene
PCP   pentachlorophenol

face difficulties in acquiring insurance coverage for cleanup of major spills, especially when some contamination already exists from minor accidental spills that have occurred over a prolonged period. Although stopping or minimizing additional pollution input has decreased pesticide detections in soil and groundwater at these sites, in most cases additional remediation is warranted (*23*). Currently, there is a need to provide dealerships with viable technologies for remediating soils and groundwater.

Site Description. We are working with an agrochemical dealership in Iowa plagued by herbicide contamination in the groundwater since the mid-1980s. Although three of the city's six drinking water wells lie within the boundaries of the site, the State of Iowa determined that minimal imminent human health hazards exist through drinking water contamination. A variety of herbicides such as atrazine, metolachlor, alachlor, and trifluralin have been detected in both surface soil and groundwater below the site.

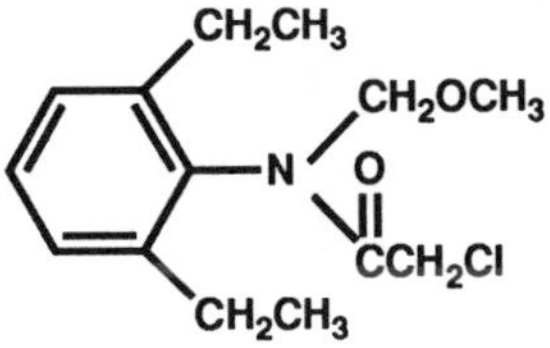

Alachlor

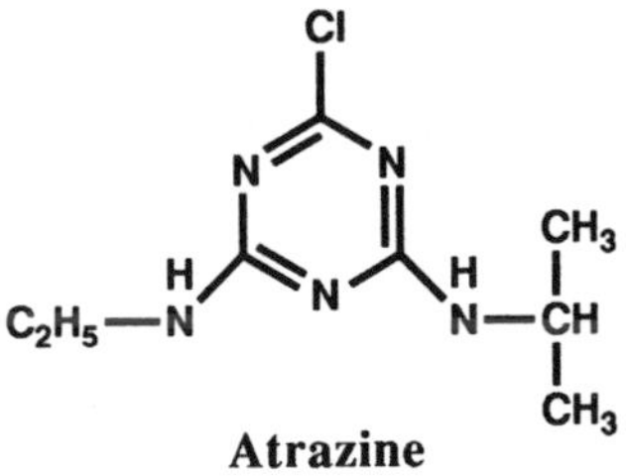

Atrazine

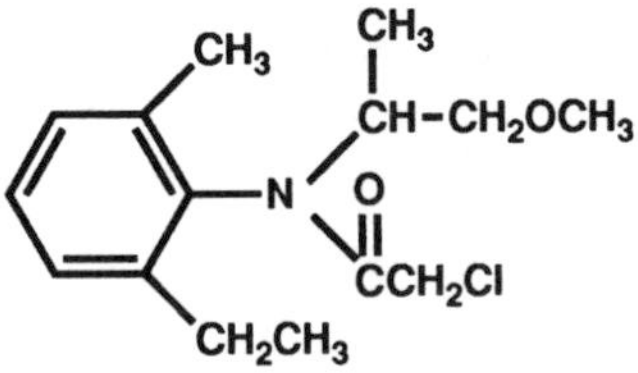

Metolachlor

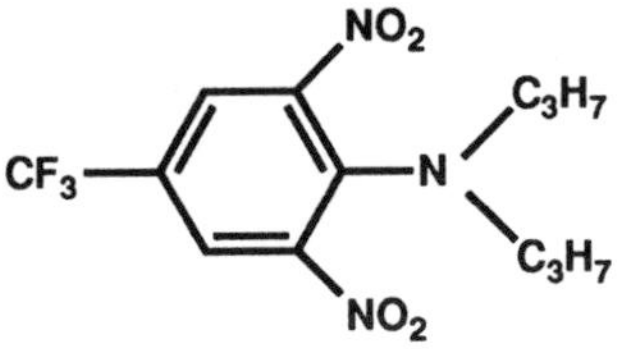

Trifluralin

In most cases, the contaminants are confined primarily to the upper 40 cm of soil. The pesticide wastes, however, are not homogeneously distributed within the site; some specific areas contain concentrations severalfold above field application rates while other areas contain concentrations less than field application rates. Physicochemical properties of a composite soil sample from the site are reported in Table 2.

Although chemical concentrations in certain samples from the site were several times field application rates as revealed by residue analyses, there was abundant plant growth. An initial vegetation survey of the site revealed several herbicide-tolerant plant biotypes including *Kochia* sp., knotweed (*Polygonum* sp.), and crabgrass (*Digitaria* sp.). That soil contaminants have been present at least 10 years provides for probable selection of microorganisms and vegetation able to survive the toxic effects of

Table 2. Physicochemical properties of a composite soil sample from nonvegetated areas[A] at the pesticide-contaminated site

Texture	Sand %	Silt %	Clay %	Organic Matter %	pH	CEC[B] (meq/100g)	Chemical Concentration/Normal Field Rate[C]		
							trifluralin	atrazine	metolachlor
sandy loam	72	16	12	2.2	7.5	12.9	0.3/0.9	0.5/0.6	9.6/1.26

[A] Soil taken from nonvegetated areas of various contamination concentrations.
[B] Cation exchange capacity
[C] Chemical concentration of the composite soil. Normal field rate is based on product label information, uniform mixing in a 15-cm soil layer and a soil density of 1.83 g/cm^3. Chemical concentrations and field rates are reported as µg/g (ppm).

the numerous agrochemicals. The presence of these plants is consistent with previous observations in the literature of their tolerance of certain herbicide classes, as well as with the types of chemicals present in the soil.

Degradation and Toxicity Studies. Experiments in the greenhouse have focused on testing whether a commodity plant such as soybean could survive in this soil and if its presence could enhance biodegradation of the three predominant chemicals–atrazine, metolachlor, and trifluralin. Although soybean survival in these soils was high, its presence did not seem to enhance the degradation of the chemicals. Because soybean is particularly sensitive to damage from these chemicals, the results are in keeping with our hypothesis that the degradation of herbicides may be enhanced only in the rhizosphere of tolerant plants. The time that these chemicals have been present in the soil has undoubtedly affected their bioavailability (and, thus, toxicity) to soybean. This is further illustrated by the fact that the addition of similar concentrations of chemicals to uncontaminated soil prohibited the growth of soybean seedlings.

Preliminary tests in the growth chamber with nonvegetated soils and rhizosphere soils collected from the root zone of *Kochia* sp. indicated enhanced degradation of the three test chemicals in the rhizosphere soil. In a second set of experiments, soils from the site were spiked with additional concentrations of the test chemicals. Although chemical concentrations in all soils were near or exceeding field application rates, these soils were spiked with additional amounts of the three chemicals so as to increase the concentrations to levels typical of point-source spills. Enhanced degradation in the *Kochia* sp. rhizosphere soil was observed in these tests (31%, 57%, 51% vs. 52%, 76%, 72% remaining for trifluralin, atrazine, and metolachlor after 14 days in rhizosphere and nonvegetated soil, respectively). Values are reported as a percentage of sterile control (autoclaved) soil. This represents a significant ($p \leq 0.10$) decrease in persistence of the herbicide mixture in the rhizosphere soil compared with the nonvegetated soil. Because plants were absent in these tests, root uptake of the herbicides was eliminated and therefore did not obscure the degradation data. However, the absence of a living plant in the rhizosphere soil during the degradation experiments may have reduced the impact of the rhizosphere microorganisms by changing the composition of the microbial community. Nonetheless, degradation of the parent compounds was significantly accelerated in the rhizosphere soil. In addition, *Kochia* sp. seedlings have emerged from the rhizosphere soils spiked with additional concentrations of the three test chemicals, indicating the ability of these plants to survive in soils containing elevated concentrations of herbicide mixtures (≥ 10 ppm).

Microbial Counts and Respiration. Estimates of microbial numbers (colony-forming units per gram of soil) in *Kochia* sp. rhizosphere soil and nonvegetated soil were made by the spread plate technique on trypticase soy agar. The rhizosphere soil had an order of magnitude higher numbers (4.2×10^5) as compared with the nonvegetated soil (3.5×10^4). In addition to the initial plate counts, determinations of colony-forming units were also conducted on pesticide-treated soils over the course of the degradation experiments. Microbial numbers increased in both soils after herbicide treatment. After 14 days, microbial numbers were 1.3×10^8 and 1.8×10^6 in rhizosphere and nonvegetated soil, respectively.

Microbial respiration, measured as carbon dioxide efflux from soils, in rhizosphere and nonvegetated soils corroborated the plate count data. Initial respiration rates in untreated *Kochia* rhizosphere soil were 1.6 to 2.4 times greater than respiration rates in nonvegetated soils, indicating more microbial biomass in the rhizosphere soil. However, differences in respiration rates between rhizosphere and edaphosphere soils decreased as the experiment progressed, suggesting that a living plant in the rhizosphere soil was necessary to maintain the elevated biomass. Respiration rates in both rhizosphere and edaphosphere soils treated with 10 ppm of the herbicide mixture were greater than respiration rates in control (untreated) soils suggesting that the herbicide mixture was not toxic to the microorganisms in the contamined site soil.

Conclusions

Microorganisms play a vital role in maintaining the environment in its current state through their metabolic activities. The highly versatile metabolic capabilities of fungi and bacteria can be used to reclaim polluted ecosystems and to minimize the potential adverse effects of hazardous chemicals released to the environment provided that a sufficient consortium of microorganisms capable of degrading the contaminant(s) are present and that environmental conditions conducive to degradation are maintained. Occasionally, environmental conditions onsite may significantly hinder microbial degradation of toxicants. In such instances, microbial degradation may be enhanced by altering conditions through nutrient additions, irrigation, tillage, or other interventions. The addition of external carbon sources may be especially important when the contaminant is degraded cometabolically.

The information presented herein illustrates the role that rhizosphere microbial communities could play in maintaining and/or remediating soil systems through metabolism of hazardous organic compounds in the root zone. The rhizosphere contains a diverse microbial community capable of vast metabolic activities. A better understanding of the complex relationship between plants, microorganisms, and chemicals in the root zone could be aided by characterizing the microbial communities associated with different plant species under contaminated and uncontaminated conditions, determining the role of exudates in selection of those communities, and investigating the role of mycorrhizae in biological degradation in the root zone.

The purpose of these experiments was not to obtain information for remediating the specific site described, but, rather, to attempt to understand why plants at these sites, in relation to the type of chemical contamination, enhance microbial degradation in the root zone. Ultimately, such information will help provide management strategies that may facilitate the biological remediation of these contaminated sites.

Acknowledgments

The authors thank Jennifer Chaplin and Pamela Rice for technical assistance. Soybean seeds used in this study were supplied by Joe Kresser. This project was partly funded by a seed grant from the Center for Health Effects of Environmental Contamination, University of Iowa. Journal Paper No. J-15478 of the Iowa Agriculture and Home Economics Experiment Station, Ames, Iowa. Project No. 3187.

Literature Cited

1. Curl, E. A.; Truelove, B. *The Rhizosphere*; Springer-Verlag: Berlin, 1986.
2. Hsu, T. S.; Bartha, R. *Appl. Environ. Microbiol.* **1979**, 37, 36-41.
3. Reddy, B. R; Sethunathan, N. *Appl. Environ. Microbiol.* **1983**, 45, 826-829.
4. Sandmann, E; Loos, M. A. *Chemosphere.* **1984**, 13, 1073-1084.
5. Lappin, H. M.; Greaves, M. P.; Slater, H. *Appl. Environ. Microbiol.* **1985**, 49, 429-433.
6. Rasolomanana, J. L.; Balandreau, J. *Rev. Ecol. Biol. Sol.* **1987**, 24, 443-457.

7. Aprill, W.; Sims, R. C. *Chemosphere.* **1990**, 20, 253-265.
8. Walton, B. T.; Anderson, T. A. *Appl. Environ. Microbiol.* **1990**, 56, 1012-1016.
9. Knaebel, D. B.; Vestal, J. R. *Can. J. Microbiol.* **1992**, 38, 643-653.
10. Shimp, J. F.; Tracy, J. C.; Davis, L. C.; Lee, E.; Huang, W.; Erickson, L. E.; Schnoor, J. L. *Crit. Rev. Environ. Sci. Technol.*, **1993**, *23*, 41-77.
11. Anderson, T. A.; White, D. C.; Walton, B. T. In *Biotransformations: Microbial Degradation of Health-Risk Compounds.* Singh, V. P.; Smith, J. E., Eds.; Rastogi and Company: Meerut, India, 1993, (Submitted).
12. Anderson, T. A.; Guthrie, E. A.; Walton, B. T. *Environ. Sci. Technol.* **1993**, 27, 2630-2636.
13. Hale, M. G.; Foy, C. L.; Shay, F. J. *Adv. Agron.* **1971**, 23, 89-109.
14. Hale, M. G.; Moore, L. D. *Adv. Agron.* **1979**, 31, 93-124.
15. Kaiser, P.; Reber, H. *Meded. Fac. Landbouwwet. Rijksuniv. (Gent.).* **1970**, 35, 689-705.
16. Herring, R.; Berring, C. L. *Bull. Environ. Contam. Toxicol.* **1988**, 40, 626-632.
17. Krueger, J. P.; Butz, R. G.; Cork, D. J. *J. Agric. Food Chem.* **1991**, 39, 1000-1003.
18. Liu, C.-M.; McLean, P. A.; Sookdeo, C. C.; Cannon, F. C. *Appl. Environ. Microbiol.* **1991**, 57, 1799-1804.
19. Gudin, C. *Proceedings of the International Symposium on Ground Water Pollution by Oil Hydrocarbons*; Prague, Czechoslovakia, 1978; pp 411-417.
20. Anderson, T. A.; Walton, B. T. *Comparative Plant Uptake and Microbial Degradation of Trichloroethylene in the Rhizospheres of Five Plant Species: Implications for Bioremediation of Contaminated Surface Soils-ORNL/TM-12017*; Oak Ridge National Laboratory: Oak Ridge, TN, 1992.
21. Walton, B. T.; Anderson, T. A.; Deckert, D. J.; Jen, M. S. *The Toxicologist, Abstracts of the 31st Annual Meeting of the Society of Toxicology*; The Society of Toxicology: Seattle, WA, 1992; p 336.
22. Myrick, C. A. In *Pesticide Waste Management: Technology and Regulation.* Bourke, J. B.; Felsot, A. S.; Gilding, T. J.; Jensen, J. K.; Seiber, J. N., Eds. American Chemical Society: Washington, DC, 1992, pp 224-233.
23. Gannon, E. *Environmental Clean-up of Fertilizer and Agri-Chemical Dealer Sites.* Iowa Natural Heritage Foundation: Des Moines, IA, 1993.
24. Buzicky, G.; Liemandt, P.; Grow, S.; Read, D. In *Pesticide Waste Management: Technology and Regulation.* Bourke, J. B.; Felsot, A. S.; Gilding, T. J.; Jensen, J. K.; Seiber, J. N., Eds. American Chemical Society: Washington, DC, 1992, pp 234-243.
25. Mandelbaum, R. T.; Wackett, L. P.; Allan, D. L. *Appl. Environ. Microbiol.* **1993**, 59, 1695-1701.
26. Sato, K. In *Interrelationships Between Microorganisms and Plants in Soil.* Vancura, V.; Kunc, F., Eds. Elsevier: New York, 1989, pp 335–342.
27. Abueva, A. A.; Bagaev, V. B. *Izv. Timiiryazev. Skh. Akad.* **1975**, 2, 127–130.
28. Seibert, K.; Fuehr, F.; Cheng, H. H. In *Theory and Practical Use of Soil Applied Herbicides Symposium.* European Weed Resource Society: Paris, 1981, pp 137–146.
29. Katayama, A.; Matsumura, F. *Environ. Toxicol. Chem.* **1993**, 12, 1059-1065.

RECEIVED December 9, 1993

Chapter 17

Plant and Microbial Establishment in Pesticide-Contaminated Soils Amended with Compost

Michael A. Cole, Xianzhong Liu, and Liu Zhang

Department of Agronomy, University of Illinois, 1102 South Goodwin, Urbana, Il 61801

Pesticide-contaminated soil (obtained from an agrichemical retail facility) was mixed with uncontaminated soil or with compost to determine the impact of compost compared to soil on plant establishment and growth, rhizosphere populations, and development of soil microbial populations and activity. Plants were established and grew well in pesticide-containing soil when consideration was given to compatibility between plant herbicide tolerance and the specific herbicide(s) present. Rhizosphere fungal and bacterial populations developed to a range of 100,000 to several billion units g^{-1} root, respectively. Soil bacterial populations were significantly higher in compost-containing mixes when compared to contaminated soil alone, while populations in soil mixes were not affected by any treatment. Fungal populations were significantly higher in planted mixes and in unplanted mixes with compost than they were in contaminated soil alone. Dehydrogenase activity was significantly higher in compost-containing mixes than in soil mixes.

Surveys of agrichemical retail sites in Wisconsin (*1*) and Illinois (*2*) demonstrated that soils at some sites contained pesticides at concentrations above recommended application rates. Because of potential contamination of groundwater and surface water by these pesticides, site owners may be required to conduct remedial activities. Such remediation can be prohibitively expensive and there is a definite need for simple, effective, and relatively inexpensive methods to degrade the pesticides as well as restore biological activity in contaminated materials. Other investigators (*3*) working with soils from similar sites found that it was difficult to enhance microbial activity and also found that pesticide degradation during attempted remediation was rather slow.

Direct land application of contaminated soils can be used as a remedial method in some cases, but this method is not possible with particular combinations of pesticides which are active against broad-leaf and grassy plants. Such application would also be prohibited if contaminants other than pesticides or if banned or restricted-use pesticides

0097–6156/94/0563–0210$08.00/0

were present. There can also be potential long-term legal liabilities associated with transport, disposal, or dispersal of soils whose contaminants have not been exhaustively identified.

Growth of plants in contaminated soils has a number of potential benefits for the remediation process. As described elsewhere in this volume, pesticide degradation in the rhizosphere can be rapid, thereby decreasing the time required for remediation. Second, plant uptake of soil water results in upward movement of water (*4*), which may reduce downward flow of contaminated water to groundwater. Third, successful plant establishment results in decreased erosional transport of contaminated soil to adjacent surface water. Finally, plant growth improves soil structure and provides organic materials which may stimulate microbial cometabolism of pesticides.

All of the soils we have obtained from contaminated agrichemical sites had one or more primary impediments to plant growth other than pesticide content. These impediments included inappropriate pH values for good plant growth, high soluble salts content, high bulk density, low organic matter content, low plant nutrient availability, low microbial activity, and the presence of phytotoxic organics such as diesel fuel. While it would be possible to deal with impediments individually on a site-by-site basis, we were also interested in developing more generic and less analytically intensive methods for remediation of these sites. The use of compost is emphasized in the present work because it has many virtues as a soil amendment resulting from its high organic matter content as well as a large and diversified microbial population.

With these considerations in mind, this work was conducted with the objectives of:

(1) Determining whether or not healthy plants could be established in soils containing a mixture of herbicides, some of which were present at several times recommended application rates,

(2) Comparing the benefits to plant growth and microbial proliferation and activity of mixing contaminated soil with uncontaminated soil or with compost derived from yard trimmings, and

(3) Comparing the rate and extent of herbicide degradation in mixes of contaminated soil with either uncontaminated soil or compost.

This paper describes results related to the first two objectives. Results of pesticide degradation studies will be published when that work has been completed.

EXPERIMENTAL

Soil and Compost Samples. Pesticide-contaminated soils were obtained during soil samplings at agrichemical retail sites in Illinois (*2*). The research reported here was conducted with material collected at Site 20. The specific sampling locations were near loading docks and other areas at which spills were more likely to occur. Cores of 8.1-cm diameter were collected to a depth of 457-cm and a composite sample of all contaminated cores was used for the present work. Contamination was restricted to the top 20 to 60 cm of material, most of which was road pack consisting of gravel, silt, and sand. The samples were passed through a 4 mm screen to remove large gravel and then used without further processing. A total of 22 pesticides were found in the combined sample when analyzed by USEPA methods 8080 and 8141 (*5,6*). Concentrations of the major compounds in the composite samples were calculated to be (in mg kg^{-1}): pendimethalin, 3.1; metolachlor, 1.5; trifluralin, 1.3; alachlor, 0.4; ethalfluralin, 0.25; chlordane, 0.2; atrazine, 0.15, and cyanazine, 0.12, for a total of about 7 mg kg^{-1} total pesticides. These analyses must be qualified by our observations that there was good control of both grassy and broadleaf weeds when the contaminated soil was diluted with uncontamined soil to give 1.5% (w/w) contaminated soil. At this dilution, the total concentration of analytically identified pesticides would be only 0.1 mg kg^{-1}, a concentration at which there should be little weed control. Since there was good weed control, the presence of an unidentified herbicide must be suspected.

Mature compost derived from yard trimmings was obtained from DK Recycling Systems, Inc., Lake Bluff, IL. The material which passed a 6-mm screen was used.

Control soil was a 50:50 (w/w) mixture of sand with Drummer silty clay loam soil obtained from a local gardener. No pesticides had been applied to the soil for four years. Material which passed a 6 mm screen was used.

Blends which contained 0, 1.5, 6, 12.5, 25, and 50% (w/w) contaminated soil and either control soil or compost were prepared and transferred into 15-cm diameter pots for greenhouse studies.

Plant Growth Procedures. Four pots of each soil mixture were planted with 6 seeds of sweet corn (*Zea mays*, cv. 'Golden Beauty') and placed in a greenhouse. Plants were watered weekly with NPK fertilizer. Four unplanted pots were treated in an identical manner. Two additional pots of soil were placed in 4° C storage as no-incubation controls. Sweet corn was chosen as the plant for these studies because of its tolerance to the major identified pesticides in the contaminated soil. Since one of the objectives of this work was to assess the impact of roots on pesticide degradation, smaller than optimally sized pots were used to ensure extensive root development in the soil mixes.

Plant Analysis. Plants were harvested at 40 d after planting and separated from soil. Roots and shoots were separated, and the roots were washed in tap water to remove adherent soil. Root samples used for rhizosphere analysis were processed within 15 min of collection. Dry weights of roots and shoots were determined by drying at 90° C to a constant mass.

Microbial Culture Media and Solutions. Buffer solution used to blend root samples for rhizosphere populations and to dilute samples for plate counts contained 0.28 g L^{-1} KH$_2$PO$_4$, 0.28 g L^{-1} K$_2$HPO$_4$, and 0.18 g L^{-1} MgSO$_4$.

Bacterial populations of soil and rhizosphere samples were assessed on glucose-tryptone agar (7), and fungal plate counts were determined on rose bengal agar (8).

Determination of Rhizosphere Microbial Populations. Approximately 1-g wet weight of roots was transferred to 30 mL of buffer and blended for 2 min in a high-speed mixer (Waring blender) operated at maximum speed. Samples were diluted and plated on glucose-tryptone agar or rose bengal agar. Plates were incubated at 30° C until colonies were large enough to count. For fungi, a 5 d incubation period was typical and 7 d incubation was typtical for bacteria. Population counts were calculated as colony forming units (cfu) g^{-1} dry weight of root. The rhizosphere:soil ratio (R:S ratio) of bacterial and fungal populations was calculated as:

R:S = $\dfrac{Mean\ population\ in\ rhizosphere\ (cfu\ g^{-1}\ root)}{Mean\ population\ in\ soil\ (cfu\ g^{-1}\ soil)}$

Determination of Soil Microbial Populations. About 2-g wet weight of soil were added to 100 mL sterile buffer and shaken at 100 rpm on a rotary shaker for 10 min. Dilutions were prepared and plated on glucose-tryptone agar (bacteria) or rose bengal agar (fungi) and incubated as described above.

Soil Dehydrogenase Activity. Dehydrogenase activity was determined by a variation of the method described by Benefield, et al. (*9*). Two grams wet weight of soil were mixed with 4.5 mL of a 1% (w/v) solution of triphenyltetrazolium chloride (TTC, Sigma Chemical Co.) and incubated at 30° C for 24 h. The red, water-insoluble formazan, which is produced by reduction of TTC, was extracted from soil by shaking with 5 mL *n*-butanol for 1 h, followed by centrifuging to separate butanol and aqueous phases. The butanol phase was removed and its absorbance at 485 nm was determined. Dehydrogenase activity is expressed as μ mol formazan produced g^{-1} dry soil 24 h^{-1}.

RESULTS AND DISCUSSION

Plant Growth. Corn plants grew well in all mixes with the exception of roots from 100% contaminated soil. In this case, the roots did not fully occupy the soil volume, were non-fibrous and displayed cortical hypertrophy. Shoot growth was normal-appearing in all cases.

As the data in Table I indicate, there were no significant differences in root mass among treatments.

Shoot production was not significantly different among mixes containing control soil + contaminated soil (hereafter referred to as "soil mixes"), as shown in Figure 1. Shoot production was significantly greater in mixes of compost + contaminated soil (hereafter referred to as "compost mixes") containing 1.5, 25, and 50% compost. These values were also significantly greater than shoot production in 100% control soil.

Total plant production (roots + shoots) was not significantly different among soil mixes, (Figure 2), but total plant production was significantly greater in compost mix containing 50% compost.

Microbial Populations in the Rhizosphere. There were no significant differences among treatments in fungal or bacterial populations of the rhizosphere (Table II). Fungal populations in the range of 10^5 cfu g^{-1} dry weight of root and bacterial populations of 10^9 cfu g^{-1} soil were similar to values reported by other researchers (*10,11*) for plants growing in uncontaminated soils. The rhizosphere:soil (R:S) ratio was calculated from the data in Table II. The ratio for fungi was 1.17:1 (soil mixes) and 1.03:1 (compost mixes). The corresponding ratios for bacteria were 45.4:1 (soil mixes) and 14.3:1 (compost mixes). These ratios are also consistent with previously published values (*10,11*). There was no significant difference between soil and compost mixes in the R:S ratio for fungi. The R:S ratio for bacteria was significantly lower for compost mixes than for soil mixes. The

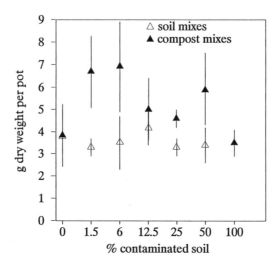

Figure 1. Shoot production by plants grown in soil or compost mixes. Bars indicate 95% confidence limits.

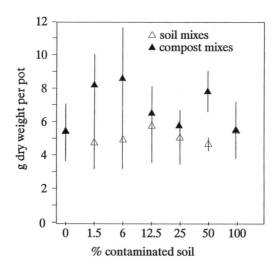

Figure 2. Total dry matter production by plants grown in soil or compost mixes. Bars indicate 95% confidence limits.

Table I. Root Production in Soil and Compost Mixes

% contaminated soil	Soil Mixes[a]	Compost Mixes[b]
	g dry weight pot^{-1}	g dry weight pot^{-1}
0	1.54 x^b	1.12 x
1.5	1.17 x	1.80 x
6	1.66 x	1.03 x
12.5	1.52 x	1.34 x
25	1.40 x	1.43 x
50	1.39 x	1.34 x
100	1.97 x	1.97 x

[a] Values are means of four pots per treatment.
[b] Values in columns followed by different letters (x,y,z) are significantly different (P = 0.05).

Table II. Microbial Populations in the Rhizosphere of Plants Grown in Soil or Compost Mixes

% contami-nated soil	Soil Mixes, Fungi	Compost Mixes, Fungi	Soil Mixes, Bacteria	Compost Mixes, Bacteria
	cfu g^{-1} mix (X 10^{-4}) [a]		cfu g^{-1} mix (X 10^{-8}) [a]	
0	20.4 x [b]	56.2 x	31.6 x	12.6 x
1.5	33.9 x	26.3 x	7.9 x	12.6 x
6	30.2 x	14.1 x	15.8 x	25.1 x
12.5	17.0 x	29.5 x	15.8 x	25.1 x
25	7.2 x	23.4 x	63.1 x	10.0 x
50	11.8 x	28.8 x	126 x	20.0 x
100	15.8 x	15.8 x	12.6 x	20.0 x

[a] Values are means of four pots per treatment.
[b] Values in columns followed by different letters (x,y,z) are significantly different (P = 0.05).

difference was the result of the much higher bacterial populations in compost mixes compared to soil mixes.

Taken together, the results indicate that corn roots growing in soil containing herbicide concentrations above label recommendations can still produce roots which sustain typical rhizosphere populations.

Microbial Populations in Soil and Compost Mixes.

Bacterial Populations in Soil and Compost Mixes. There were no significant differences in bacterial populations in any of the soil mix treatments (Table III). Populations in planted compost mixes were not significantly greater than 100% contaminated soil. Values for compost mixes that were unplanted or stored at 4° C were significantly greater than 100% contaminated soil. Bacterial populations were significantly higher in nearly all compost mixes when compared to soil mixes. What is not evident from the population values is that microbial diversity, as indicated by the variety of colony types appearing on dilution plates, was much higher in all mixes than it was in the contaminated soil. Contaminated soil had a very large population of a single colony type for bacteria, in contrast to the mixes with 10 to 20 colony types.

Between-sample variability was very high in compost mixes and in planted treatments. This variability is probably was the result of slow bacterial movement from the compost into adjacent soil. In such a situation, a single sample might contain a larger amount of compost with a very high bacterial population and relatively little soil with a smaller population, whereas the next subsample might contain more soil and less compost and have a smaller overall population. We have encountered similar variability in microbial populations and activity when sampling recently reconstructed soils following surface mining (unpublished results).

Fungal Populations in Soil and Compost Mixes. There were no significant differences in fungal populations in soil mixes with or without plants (Table IV), while most values for samples stored at 4° C were significantly greater than 100% contaminated soil. Fungal populations in 100% contaminated soil with plants was significantly greater than observed in either unplanted soil or in 4° C control. The results demonstrated that dilution of contaminated soil had a beneficial effect on fungal proliferation, probably due to the reduced toxicity of the diluted soil mixes.

Compost mixes, which had not been planted or which were stored at 4° C, had significantly higher fungal populations than 100% contaminated soil (Table IV). There were no significant differences among compost mixes in which corn had grown (Table IV).

Comparison of fungal populations in unplanted soil mixes with unplanted compost mixes indicates that compost addition (Table IV) resulted in an approximately tenfold greater fungal population than seen with soil mixes (Table IV). Fungal diversity was also low in contaminated soil with only two species growing on the dilution plates, in contrast to 10 to 15 recognizably different fungal species in mixes containing control soil or compost.

Table III. Bacterial Populations in Soil and Compost Mixes

% contaminated soil	Soil Mixes			Compost Mixes		
	Planted	Not Planted	4°C Storage	Planted	Not Planted	4°C Storage
	cfu g⁻¹ dry weight of mix (× 10⁻⁶)[a]			cfu g⁻¹ dry weight of mix (× 10⁻⁶)[a]		
0	91.2 x[b]	20.0 x	38.0 x	179 x	417 y	1230 y
1.5	75.9 x	55.0 x	11.2 x	246 x	269 y	1148 y
6	42.7 x	20.9 x	46.8 x	105 x	355 y	813 y
12.5	56.2 x	39.8 x	15.5 x	71 x	142 y	661 y
25	56.2 x	56.2 x	26.3 x	162 x	339 y	126 y
50	33.1 x	40.7 x	17.8 x	85 x	115 y	295 y
100	51.3 x	18.6 x	22.9 x	51 x	19 x	23 x

[a] Values are means of four pots each for planted and not planted treatments and two pots for 4° C storage.
[b] Values in columns followed by different letters (x, y, z) are significantly different (P = 0.05).

Table IV. Fungal Populations in Soil and Compost Mixes

% contaminated soil	Soil Mixes			Compost Mixes		
	Planted	Not Planted	4°C Storage	Planted	Not Planted	4°C Storage
	cfu g^{-1} dry	weight of soil	(X 10^{-4})[a]	cfu g^{-1} dry	weight of soil	(X 10^{-4})[a]
0	6.8 x[b]	2.1 x	1.8 y	60 x	155 y	182 y
1.5	1.6 x	1.1 x	2.5 y	38 x	36 y	162 y
6	4.5 x	2.2 x	1.1 x	22 x	29 y	32 y
12.5	5.1 x	2.8 x	1.7 y	20 x	37 y	21 y
25	1.7 x	2.6 x	1.9 y	17 x	20 y	22 y
50	1.4 x	3.0 x	2.5 y	12 x	24 y	27 y
100	4.3 x	0.6 x	0.3 x	4.3 x	0.6 x	0.3 x

a Values are means of four pots each for planted and not planted treatments and two pots for $4o$ C storage.
b Values in columns followed by different letters (x,y,z) are significantly different (P = 0.05).

Dehydrogenase Activity in Soil and Compost Mixes.

Dehydrogenase Activity in Soil Mixes. Dehydrogenase activity has been used by numerous investigators as an indication of overall microbial activity in soil (*12*). There were no significant differences among treatments in dehydrogenase activity in soil mixes, although differences among treatments were substantial. Variability of results within treatments was large with this assay. Other researchers who conducted soil restoration activities following surface mining have encountered similar variability problems (*13*).

Dehydrogenase Activity in Compost Mixes. Dehydrogenase activity was significantly greater in all compost-containing mixes when compared to 100% contaminated soil (Table V). Activity was significantly lower in planted and unplanted mixes containing 50% or 75% contaminated soil.

Dehydrogenase values were lower than would be predicted on the basis of no interaction between contaminated soil and compost. For example, the predicted activity in mixes containing 25% contaminated soil + 75% compost would be about 1100   mol product g^{-1} mix. This value was obtained by formula:

$$Total\ activity = (Activity\ of\ soil\ x\ fraction\ of\ soil)$$
$$+ (Activity\ of\ compost\ x\ fraction\ of\ compost).$$

The expected value for 25% contaminated soil with plants would be:

$$Total\ activity = (18\ x\ 0.25) + (1464\ x\ 0.75) = 1102\ \ \ mol\ product.$$

The observed activity was approximately one-half of the predicted activity, which suggested that the contaminated soil had significant inhibitory effects on microbial activity. In spite of this inhibition, microbial activity was still 10- to 20- fold higher in compost-containing mixes than in soil mixes (Table V). These results indicated that compost had a dramatic, positive impact on microbial activity in mixes containing contaminated soils. Preliminary results showed that degradation of some pesticides is also stimulated when compost rather than uncontaminated soil is mixed with contaminated soil.

CONCLUSIONS

Addition of compost to pesticide-contaminated soil significantly increased plant dry matter production.   No attempt was made to identify specific reasons for plant growth stimulation by the compost; this phemenon is commonly reported (*14*). Compost contains water-soluble trace elements (Cole, unpublished data) which may have been unavailable in the contaminated soil and it also improved soil physical properties by reducing bulk density. Adsorption of pesticides by the organic matter in compost may have reduced phytotoxicity as well.

Microbial populations and activity are often reduced in xenobiotic-impacted soil and introduction of exogenous organisms produced in the laboratory has not been very

Table V. Dehydrogenase Activity in Soil and Compost Mixes

% contaminated soil	Soil Mixes			Compost Mixes		
	Planted	Not Planted	4° C Storage	Planted	Not Planted	4° C Storage
	μmol	formazan	g^{-1} mix 24 h^{-1a}	μmol	formazan	g^{-1} mix 24 h^{-1}
0	68 x^b	40 x	54 y	1464 x	1299 x	1213 x,y
1.5	67 x	43 x	57 y	1080 x,y	703 x	1621 x
6	50 x	24 x	50 y	1075 x	972 x	1222 y
12.5	54 x	33 x	54 y	1052 x	825 x	1396 y
25	59 x	25 x	38 x	575 y	613 y	1052 x
50	32 x	25 x	21 x	370 y	336 y	324 z
100	18 x	16 x	12 x	18 z	16 z	12 w

[a] Values are means of four pots each for planted and not planted treatments and two pots for 4° C storage.
[b] Values in columns followed by different letters (w,x,y,z) are significantly different ($P = 0.05$).

successful. In contrast, survival of bacteria and fungi added in compost was very good. Bacterial and fungal populations in compost mixes were several-fold higher than populations in soil mixes. Although bacterial populations were lower in unplanted mixes when compared to 4° C storage (Table III), they were still higher than seen when uncontaminated soil was used as an inoculum. Fungal populations in unplanted soil were similar to values for 4° C storage, which suggests that compost is a particularly good inoculum for introduction of fungi into soil. Successful introduction of fungi with biocontrol capabilities using compost as an inoculum has been demonstrated (*15*).

It should be noted that, as a result of potential inhibition of microbial activity by contaminated soil, addition of the relatively small amounts of compost and other organic materials that are more typically used as a soil amendment (about 20 to 40 tons hectare^{-1}) is not likely to have the large beneficial effect on microbial populations and activity that addition of high rates of compost have.

ACKNOWLEDGMENTS

This work was supported by a grant from the Illinois Department of Energy and Natural Resources, Hazardous Waste Research and Information Center, and funds from Solum Remediation Services, Lake Bluff, IL.

The assistance of Warren Goetsch, Illinois Department of Agriculture, in obtaining soil samples is gratefully acknowledged.

LITERATURE CITED

1.   Habecker, M.A. *Environmental Contamination at Wisconsin Pesticide Mixing/ loading Facilities: Case Study, Investigation and Remedial Action Evaluation.* Wisconsin Dept. of Agriculture, Trade and Consumer Protection, 1989.
2.   Krapac, I.G.; Roy; W.R., Smyth; C.A., Barnhardt, M.L. *Occurrence and Distribution of Pesticides in Soil at Agrichemical Facilities in Illinois.* Final Report to Illinois Department of Agriculture, 1993.
3.   Felsot, A.; Dzantor, E.K.; Case, L.; Liebl, R. *Assessment of Problems Associated with Landfilling or Land Application of Pesticide Waste and Feasibility of Cleanup by Microbial Degradation.* Report No. HWRIC RR-053, Illinois Department of Energy and Natural Resources, Springfield, IL.
4.   Hillel, D. Applications of Soil Physics; Academic Press: New York, pp 169-196.
5.   USEPA. *Test Methods for Evaluating Solid Waste*, SW-846, U.S. Environmental Protection Agency, Washington, DC, 1986.
6.   USEPA. *Test Methods for Evaluating Solid Waste. Physical/Chemical Methods-Third Ed. Proposed Update Package.* U.S. Environmental Protection Agency, Washington, DC, 1989.
7.   Cole, M.A.; Turgeon, A.J. *Soil Biol. Biochem.*,1978,*10*,181.
8.   Martin, J.P. *Soil Sci.*,1950,215.
9.   Benefield, C.B.;Howard, P.J.A.; Howard, D.M. *Soil Biol. Biochem.*,1977, *11*,67.

10. Rovira, A.D.; Davey, C.B. In *The Plant Root and Its Environment*; Univ. Press of Virginia, Charlottesville, VA, 1974, pp 153-204.
11. Moorman, T.B.; Dowler, C.C. *Agric. Ecosys. Environ.* **1991**,*35*, 311.
12. Schaffer, A. In *Soil Biochemistry*, volume 8, Bollag, J-M.; Stotzky, G. Eds., Marcel Dekker: New York, 1993, pp 273-340.
13. Lindemann, W.C.; Lindsey, D.L.; Fresquez, P.R. *Soil Sci. Soc. Amer. J.*, **1984**,*48*, 574.
14. Chanyasak, V.; Katayama, A.; Hirai, M.F.; Mori, S.; Kubota, H. *Soil Sci. Plant Nutr*,.**1983**,*29*, 239.
15. Hoitink, H.A.J.; Boehm, M.J.; Hadar, Y. In *Science and Engineering of Composting*; Renaissance Publications, Worthington, OH, 1993, pp 601-621.

RECEIVED February 16, 1994

INDEXES

Author Index

Anderson, Todd A., 82,199
Balandreau, Jacques, 28
Banks, M. K., 132
Barkovskii, Andrei L., 28
Bollag, Jean-Marc, 2,102
Boullant, Marie-Louise, 28
Boyle, J. J., 70
Chaffin, C., 112
Christman, Russell F., 82
Coats, Joel R., 199
Cole, Michael A., 210
Davis, Lawrence C., 112
Dec, Jerzy, 102
Donnelly, Paula K., 93
Engelke, M. C., 142
Erickson, L. E., 112
Fateley, W. G., 112
Fletcher, John S., 93
Guthrie, Elizabeth A., 11,82
Hammaker, R. M., 112
Hoagland, Robert E., 160,184
Hoylman, Anne M., 11,82
Johnson, Theodore R., 82

Kendall, E. W., 142
Knaebel, David B., 56
Kruger, Ellen L., 199
Liu, Xianzhong, 210
Locke, Martin A., 160,184
Mertz, Tawna, 2
Muralidharan, N., 112
Otjen, Lewis, 2
Perez, Mary M., 82
Qiu, X., 142
Sato, Kyo, 43
Schwab, A. P., 132
Shah, S. I., 142
Shann, Jodi R., 70
Sims, R. C., 123,142
Sorensen, D. L., 123,142
Vestal, J. Robie, 56
Visser, V. P., 112
Walton, Barbara T., 11,82
Watkins, J. W., 123
Zablotowicz, Robert M., 160,184
Zhang, Liu, 210

Affiliation Index

Agricultural Research Service, 160,184
Iowa State University, 82,199
Kansas State University, 112,132
Oak Ridge National Laboratory, 11,82
Pennsylvania State University, 2,102
St. Olaf College, 82
Texas A&M University, 82,142
Tohoku University, 43
Union Carbide Corporation, 142

Université Claude Bernard-Lyon, 28
University of Cincinnati, 56,70
University of Idaho, 56
University of Illinois, 210
University of North Carolina, 11,82
University of Oklahoma, 93
University of Tennessee, 11,82
U. S. Department of Agriculture, 160,184
Utah State University, 123,142

Subject Index

A

Acenaphthene, in soil, 152,153,154t
Acenaphthylene, in soil, 153,154t
Acetagens, anaerobic degradation of
    hazardous chemicals, 21
Acetochlor, metabolism by glutathione
    conjugation, 185
Acetylene-reduction activity
    Azospirillum lipoferum, 35–38
    effect of caffeic acid, 39
    electron acceptor-limiting conditions,
        35*f*
    measurement, 29–31
    with O_2 and caffeic acid, 36*f*
Achromobacter sp., adaptation to high
    root exudate levels, 13
Acids, contaminant, 3
Acifluorifen, metabolism by glutathione
    conjugation, 185
Acinetobacter, aerobic degradation of
    phenolics, 29
Actinomycetes
    associations with vascular plants, 14t
    description and location, 15t
    oxidoreductase activity, 103
    recycling of soil organic matter, 2
Additives, enhancers of biodegradation, 71
Adsorption, occurrences in rhizosphere, 76
Aerobacter, description and location, 15t
Aerobic microorganisms, microhabitats, 16
Aerobic mineralization, organic matter, 4
Aerobic vs. anaerobic metabolism of
    toxicants, 16–21
African clover, sensitivity to 2,4-D, 203t
Agaricales fungi, tyrosinase activity, 103
Agricultural dealerships, contaminants, 205
Agricultural soil, degradation, 56–69
Agricultural wastes, biological treatment, 3
Agrobacterium sp., adaptation to high root
    exudate levels, 13
Agrochemical(s), bioremediation, 19

Agrochemical dealership
    contamination of soil and water, 202,205
    pesticide wastes, 199–209
Agropyro christatum, PAH degradation, 144
Agropyron smithii, PAH degradation, 144
Alachlor
    aquachemical dealership, 205
    metabolism
        by cell suspension, 185–188
        glutathione conjugation, 185
        rhizobacteria, 185–188,193-194
    occurrence in soil, 212
Alachlor-γ-glutamyltranspeptidase activity
    and catabolism by cell-free extracts, 191t
Alachlor metabolic products,
    186,189*f*,190*f*,192*f*
Alcaligenes faecalis, metabolism of
    3-chloroaniline, 163
Alcaligenes spp.
    aerobic degradation of phenolics, 29
    propanil aryl acylamidase activity, 169t
Alfalfa
    adsorption of naphthalene, 135–136
    biomass of roots and shoots, 135t
    effect on biodegradation and
        volatilization, 112–122
    effect on naphthalene volatilization, 19
Alkylbenzene sulfonates, mineralization, 4
Allelopathic compounds, degradation by
    rhizosphere microbiota, 16
Amino acid(s)
    effect on binding matrix, 68
    effect on PCP degradation, 43
    in organic rhizosphere exudates, 72
    transformation to nitrate via ammonium
        and nitrite, 46
Amino acid-treated soil, effect of PCP on
    bacterial population, 46,47*f*
Ammonification, effect of PCP, 46
Anaerobic degradation system
    bacteria involved, 21
    phenol degradation, 117

Anaerobic environments, electron
  acceptors, 4
Anaerobic microorganisms
  microhabitats in rhizosphere, 16
  role in degradation of pollutants, 4
Andropogon gerardi, PAH degradation, 144
Angiospermae
  endomycorrhizae, 17t
  nitrogen-fixing prokaryotes, 14t
Aniline(s), cleanup by bioremediation, 6
Aniline-related wastes, generation of
  phenolic compounds, 28
Animals, biological defenses against
  toxicants, 83
Anionic surfactants, biodegradation, 56
Annual plants, roots, 13
Anthracene
  concentration in soil, 153,154t
  resistance to degradation, 134
  uptake and accumulation in plants, 133
Anthropogenic pollutants
  decomposition by fungal enzymes, 19
  degradation by rhizosphere microbiota, 16
  examples, 19
Antibiotics, role of root exudates, 75
Aqueous phase organics, analysis, 115
Aromatic compounds
  binding to humic substances, 7
  degradation by ectomycorrhizal fungi, 94
  degradation by mycorrhizal fungi, 95
Aromatic hydrocarbons
  degradation by bacteria, 18t
  metabolism by prokaryotes, 19
Aromatic ring
  bacterial metabolism, 19
  metabolism in plant tissues, 143–144
Arthrobacter sp.
  degradation of pentachlorophenol, 179
  high root exudate levels, 13
  peroxidase activity, 103
  propanil aryl acylamidase activity, 169t
  use as inocula for bioremediation, 6
Aryl acylamidases
  assay, 166
  propanil degradation, 161

Aryl acylamidases—*Continued*
  rhizosphere microflora, 160
Aryl amidases, inhibition by insecticides,
  180
Aspergillus spp.
  laccase, peroxidase, and tyrosinase
    activity, 103
  release of 3,4-dichloroaniline
    from humic complexes, 163
Atrazine
  aquachemical dealership, 205
  degradation by fungi, 96
  degradation pathway, 195
  effect on ectomycorrhizae, 96
  enhanced degradation in soil, 207
  metabolism
    by fungi, 93
    by glutathione conjugation, 185
    in presence of corn, 204t
    mineralization by grasses, 203t
  pesticide-contaminated site, 206t,212
  plant-assisted bioremediation, 113
Atrazine–GST studies, 195
Azospirillum lipoferum
  acetylene-reduction activity, 35–38
  adaptation to rice rhizosphere, 40
  caffeic acid as carbon source, 38–39
  culture conditions, 30
  effect of phenolics, 28,29,31–32,38
  effect on caffeic acid metabolism,
    33–35
  effect on polyphenolics, 40
  effect on UV spectra of phenolics, 37t
  enzymatic activity of cells and
    cell-free extracts, 37–38
  motile vs. nonmotile, 38
  nitrogenase activity, 35–38
  redox potential after addition of O_2 and
    caffeic acid, 36f
  role in transformation of polyphenolic
    substances, 19
  strain 4B vs. strain 4T, 30,38
Azospirillum spp.
  catabolism of phenolics, 28–29
  description and location, 15t

Azospirillum spp.—*Continued*
  oxygen conditions to express nitrogenase
    activity, 29
  root-growth-promoting activity, 28
Azotobacteraceae, description and
    location, 15t

B

Bacillus spp.
  peroxidase activity, 103
  propanil aryl acylamidase activity, 169t
Bacteria
  biological defenses against toxicants, 83
  catabolism of organic substances, 12
  detoxication of hazardous compounds, 11
  found in rhizosphere, 13
  in soil and compost mixes, 216,217t
  oxidoreductase activity, 103
  recycling of soil organic matter, 2
Bacterial inoculants for bioremediation,
    factors limiting use, 196
Barley, transpiration stream concentration
    factor values, 113
Bases, contaminant, 3
Bell Rhodesgrass, role in volatilization
    of naphthalene, 131
Benthiocarb, effect on heterotrophic
    bacteria in the rhizosphere, 203t
Benz[*a*]anthracene, effect of grasses on
    degradation, 123,134,144
Benzene
  bioremediation in the rhizosphere, 19
  degradation by anaerobic microorganisms, 4
  use of plant-assisted bioremediation, 113
Benzo[*a*]pyrene
  effect of grasses on degradation,
    123,134,144
  metabolism in plant tissues,
    133,143–144
Benzoate
  catabolism by *Azospirillum,* 29
  mineralization in soil exposed to dicot
    and monocot root exudates, 74f

Benzoate—*Continued*
  reductive degradation by
    nitrate-respiring bacteria, 22f
Bermuda grass, description, 150
Bioavailability
  definition, 76
  effect of soil organic matter, 13
  role in rhizosphere degradation, 76–79
Biochemical detoxication, defense against
    toxicants, 83
Biodegradation
  enhancement by landfarming, 6
  fate of chemical contaminants, 57
  mechanism affecting fate of naphthalene, 124
  promotion by alfalfa plants and
    associated microorganisms, 112–122
  rhizosphere microbial communities, 56–69
  role of plants, 57
Bioreactors
  biological treatment of wastes, 3
  byproducts, 5–6
  characteristics, 5
  use in soil remediation, 70
Bioreclamation, definition, 2
Bioremediation
  advantages and disadvantages, 3,8
  definition, 2
  design of laboratory studies, 91
  factors for success, 179
  future considerations, 8–9
  mycorrhizal fungi, 93–99
  Phanerochaete chrysosporium, 95
  plant–microbe–toxicant interactions, 91
  problems in implementation, 8
  role of fungi, 94
  role of microorganisms, 2–10
  role of plants and plant roots, 13,113
  techniques and approaches, 3,5–7
Biotransformation, defense against
    toxicants, 83,86f
Birds, biochemical defense against
    toxicants, 83
Birnessite, use in oxidative coupling of
    chlorophenols, 104,105t
Bluegrama, seeded grass, 151

Bouteloua species, effect on PAH
  degradation, 144
Bradyrhizobium
  degradation of aromatic and
    haloaromatic compounds, 28
  nitrogen fixation, 28
  propanil aryl acylamidase activity, 169*t*
Brines, contaminant, 3
Bromacil
  effect of transpiration rate, 114
  mechanism of degradation, 114
Brown-rot fungi, tyrosinase activity, 103
Buchloe dactyloides, in experiments on
  vegetative bioremediation, 145
Bush bean, mineralization of diazinon and
  parathion, 203*t*

C

^{14}C, amount bound to vegetated,
  unvegetated, and poisoned
  microcosms, 130
^{14}C acetate, measurement of soil
  microbial activity, 60
^{14}C chemicals
  mineralization results, 65,66*f*,67*f*
  structures, 59*f*
  uptake by *M. alba*, 87
  use in biodegradation of surfactants
    study, 58
^{14}C substrates, biodegradation studies, 57
^{14}C surfactants, utilization by
  microorganisms, 68
^{14}C uptake studies, methods, 88–89
Cadmium, levels at mine spoil sites, 96
Caffeic acid
  analysis of metabolites by HPLC, 33–35
  analysis of metabolites by TLC, 32–33
  effect on *A. lipoferum,* 28–42
  effect on acetylene-reduction activity, 39
  extraction and characterization of
    metabolites, 31
  incubation with *A. lipoferum* enzymes,
    37–38

Caffeic acid—*Continued*
  metabolites, 38–39
  redox potential changes, 39
  UV spectra after reaction with *A.
    lipoferum* cells or cell-free extract, 37*t*
Carbamate insecticides, inhibition of aryl
  amidase activity, 180
Carbohydrates, degradation by cell-free
  enzymes, 7
Carbon compounds, role in xenobiotic
  degradation, 71
Carbon dependence and sources,
  mycchorizal fungi, 95
Carbon dioxide, monitoring in soil, 148
Carbon in root exudates, source, 72
Carboxypeptidase, in glutathione
  conjugate metabolism, 188
Catabolic enzymes, from mycorrhizal
  fungi, 94–95
Catechol
  effect on growth of *A. lipoferum*, 31,38
  incubation with *A. lipoferum* enzymes,
    37–38
UV spectra after reaction with *A.
  lipoferum* cells or cell-free extract, 37*t*
Cationic surfactants, biodegradation, 56
CDNB–GST (1-chloro-2,4-dinitroben
  zene–glutathione *S*-transferase),
  activity in rhizosphere enrichments,
    193–194
Cell-free enzymes
  preparation, 166
  use in bioremediation, 7
Cell walls, role in xenobiotic adsorption,
  76–77
Cellulase, in mat structures formed by
  ectomycorrhizae, 95
Cellulose
  decomposition by fungal enzymes, 19
  degradation mechanism, 75
  PCP degradation, 43,53–54*f*
Cenococcum geophilum, effect on
  hexazinone, 97
Cenococcum graniforme, catabolic
  enzymes, 94

Chemical plant factors in the rhizosphere, 72–75

Chemical wastes, stress to plants, 90

Chemoattractants, role of root exudates, 75

Chemotactic communication, in formation of nitrogen-fixing nodules, 16

Chlordane, occurrence in pesticide-contaminated soil, 212

Chloride, release during oxidative coupling processes, 103,109–110

Chlorinated aliphatics, bioremediation in the rhizosphere, 19

Chlorinated aromatic compounds
degradation by *P. chrysosporium,* 95
metabolism by fungi, 93
optimum conditions for degradation, 95

Chlorinated hydrocarbons,
cleanup by bioremediation, 6
mineralization, 4

Chlorinated phenols
binding to humic acid, 103
coupling reactions, catalysis by oxidoreductive enzymes, 102
dehalogenation during binding to humus, 102–111
mechanism of binding to organic matter, 102–103
polymerization mediated by birnessite, 104

Chlorinated solvents
bioremediation in the rhizosphere, 19
chemical stress to plants, 90

Chlorine, removal by microorganisms, 4

Chloris gayanaprairie, effect on PAH degradation, 144

Chloroacetamides
biodegradation pathways, 194
metabolism by glutathione conjugation, 185

3-Chloroanilines
propanil metabolite, 163
release of chloride ions, 107

Chlorobiphenyls, microbial substrate, 4

1-Chloro-2,4-dinitrobenzene,
nonherbicide substrate for glutathione *S*-transferase, 185

Chloronitrophenols, microbial substrate, 4

Chlorophenol(s)
binding to humic acid, 105*t*,109
coupling of free radicals, 106*f*,107–108
coupling reactions, dependence on free radical number, 108
dehalogenation numbers, 108
dimer formed by coupling, 108–109
limited dehalogenation, 107
polymerization with horseradish peroxidase vs. birnessite, 105*t*
tyrasine-mediated oxidation, mechanism, 107

Chlorpropham, degradation by ectomycorrhizzal fungi, 97

Chrysene, effect of grasses on degradation, 123,134,144

Citrobacter diversus, alachlor metabolism, 186

Citrobacter spp., propanil aryl acylamidase activity, 169*t*

Clay soils, bioremediation for PAH contamination, 142–157

$^{14}CO_2$, amount trapped from vegetated and unvegetated microcosms, 129*f*

CO_2 levels, results in soil and rhizosphere study, 120

Coal-conversion wastewaters, treatment with horseradish peroxidase, 7

Coenzyme F430, reductive dechlorination of TCE, 23*f*

Coevolution, plant–microbial system in legumes, 75

Cometabolism
definition, 21
role in mineralization, 4

Commensal relationships, role in mineralization, 4

Complexation, bioremediation, 3

Compost
effect on plant growth in pesticide-contaminated soil, 210–222
use as soil amendment, 211
use in pollution control, 6

Concentration factor, definition, 143

Conjugation, method of gene transfer, 4

Conjugation of xenobiotics, plant adsorption, 77

Conservative tracer, use in measurement of gas phase concentration, 120

Constant capacitance model, wall adsorption by maize and soybean, 77

Contaminant(s)
constituents, parameter in vegetative bioremediation, 148,149t
degradation, competition between plant and microorganism uptake, 114
dissipation by plants, 133
examples, 3
migration, determination, 148

Contaminated environments, 3

Contaminated soil, cleanup methods, 93

Copolymerization of toxicants, rhizosphere microorganisms and humic substances, 13

Corn
growth in pesticide-contaminated soil, 213,216
sensitivity to atrazine, 204t
use in biodegradation study, 56,60

Corn treatment, effect on microbial activity, 64

Coupling reactions
dependence on number of free radicals, 108
free radicals from 2,4-dichlorophenol, 106f

Crabgrass, tolerance for herbicides, 205

Cunninghamella elegans, metabolism of PAHs, 19

Cyanazine, occurrence in pesticide-contaminated soil, 212

Cyanobacteriales, associations with vascular plants, 14t

Cycadales, associations with nitrogen-fixing prokaryotes, 14t

Cysteine-β-lyases
activity in rhizobacteria, 191,193t
in glutathione conjugate metabolism, 188,191,193t

Cytochrome P–450 mixed function oxidase (MFO) system
excretion of lipophilic organic substrates, 84f
mechanism of detoxication in animals, 83,84f

D

DDT
degradation by anaerobic microorganisms, 4
degradation by rhizosphere-competent fungus, 204t

Decomposition processes, micro- and mesofauna, 12

Deep fibrous root zone, effects on PAH degradation, 143

Defense against toxicants, forms, 83

Degradation, effect of plants, 11,70–81

Degradation of environmentally persistent chemicals by anaerobic microorganisms, 4

Degradation of PCP, effect of nutrient concentration, 53

Degradation of toxicants, resultant detoxication, 11

Degradation rates, stimulation by phytoremediation, 2

Dehalogenation
chlorinated phenols, immediate cause, 103,104
chlorophenols during oxidative coupling processes, 105t
role in decomposition of organic compounds, 109

Dehalogenation number (DN)
binding to humic acid vs. polymerization, 107–108
chlorophenols, 105t
definition, 104

Dehydrogenase
activity in soil and compost mixes, 219,220t
calculation of total activity, 219
indicator of microbial activity, 219

Detoxication
factors, 11

Detoxication—*Continued*
  hypothetical mechanism, 86*f*
  in plants, 84
Diazinon
  accelerated degradation in rhizosphere,
    184,202
  effects of plants on degradation rates, 123
  mineralization in bush bean rhizosphere,
    203*t*
Dibenz[*a,h*]anthracene, effect of grasses
  on degradation, 123,134,144
Dicamba, reversal of toxic effects by soil
  microorganisms, 201
3,4-Dichloroaniline
  conjugation pathways, 179
  fate in soil, 163
  metabolic fate in plants, 161
2,4-Dichlorophenol (2,4-DCP)
  coupling of free radicals, 106*f*,107–108
  dehalogenation due to binding to humic
    acid, 107
  free radical dimerization, 104
  incorporation into humic acid, 76
  soil adsorption, effect on
    degradation, 76
2,4-Dichlorophenoxyacetic acid (2,4-D)
  carbon source for microbes in wheat
    root, 203*t*
  concentration in hydroponic solutions, 78*f*
  degradation
    availability of xenobiotics, 79
    by fungi, 93,96
    effect of soil adsorption, 76
    in sugarcane rhisozphere, 203*t*
  effect on bacteria in rhizosphere, 203*t*
  mineralization
    experimental methodology and
      results, 77
    in soil amended with wheat cell
      wall, 78*f*
    in soil exposed to dicot and monocot
      root exudates, 74*f*
Dicots, results of mineralization, 72
Dicotyledons
  2,4-D and 2,4,5-T degradation rates, 13

Dicotyledons—*Continued*
  roots, 13
  soil requirements, 13
Dieldrin, degradation by rhizosphere-
  competent fungus, 204*t*
Digitaria sp., tolerance for herbicides, 205
Dihalogenated aromatics, dechlorination
  under anaerobic conditions, 163
3,4-Dihydroxybenzoic acid
  effect of *A. lipoferum*, 31,38,40
  incubation with *A. lipoferum* enzymes,
    37–38
  UV spectra after reaction with *A.
    lipoferum* cells or cell-free extract, 37*t*
Dimer formed by chlorophenol coupling,
  108–109
Dioxygenase enzymes, role in bacterial
  detoxication, 83
Diphenyl ethers, metabolism by
  glutathione conjugation, 185
Donnan system, use as cell wall model, 77

E

Earthworms, recycling of soil organic
  matter, 2
Ectendotrophic mycorrhizae, growth in
  axenic culture, 94
Ectomycorrhizae, associations with
  vascular plants, 17*t*
Ectomycorrhizal fungi
  bioremediation agents, 93
  growth in axenic culture, 94
  lignin mineralization, 95
Ectotrophic mycorrhizae, formation of
  hyphal network, 94
Electron acceptors
  anaerobic environments, 4
  metabolism of phenolics, 29
  metabolism of polyphenolics, 40
  phenolics, 40
Electron donors, phenolics, 40
Electron-transfer properties, phenolics, 40

Elymus canadensis effect on PAH degradation, 144
Endomycorrhizae
association with plant roots, 76
association with vascular plants, 17*t*
Endospore-forming bacteria, description and location, 15*t*
Endosulfan, degradation by rhizosphere-competent fungus, 204*t*
Endotrophic mycorrhizae, penetration of host cells, 94
Enterobacter cloacae, alachlor metabolism, 186,187*f*
Enterobacter spp.,
description and location, 15*t*
propanil aryl acylamidase activity, 169*t*
Enterobacteriaceae strains, alachlor glutathione *S*-transferase, 185
Environmental conditions, effects on detoxication, 11
Environmental impact of pesticides, significance of pesticide-tolerant bacteria, 50
Environmental pollutants, effect of mycorrhizal fungi, 96–97
Environmental Protection Agency, promotion of bioremediation, 8
Environmentally persistent chemicals, degradation by anaerobic microorganisms, 4
Enzymatic electron-transfer system, phenolics, 40
Enzymatic oxidation, reduction, and hydrolysis, use by plants in detoxication, 84
Enzyme activity, measurement, 31
Enzyme immobilization, use in bioremediation, 7
Ericaceous mycorrhizae, growth in axenic culture, 94
Ericales, associations with endomycorrhizae, 17*t*
Ericoid mycorrhizal fungi, lignin mineralization, 95

Escherichia coli, use of 3,4-dichloroaniline as substrate, 163
Ethalfluralin, occurrence in pesticide-contaminated soil, 212
Eubacteriales, associations with vascular plants, 14*t*
Eukaryotes, metabolism of PAHs, 19
Eukaryotic oxidative pathways, mineralization of polycyclic aromatic hydrocarbons, 22*f*
Evaporation vs. transpiration, effect on phenol concentration, 117
Excretion, defense against toxicants, 83
Extraction–treatment techniques of bioremediation, 5–6
Exxon Valdez oil spill cleanup, results of in situ remediation, 70–71

F

Fatty acids, in organic rhizosphere exudates, 72
Fe-deficient monocot and dicot roots
accumulation of organic acids, 75
proton release, 75
Fertilizers, effect in agricultural soils, 56
Fescue
adsorption of naphthalene, 135–136
biomass of roots and shoots, 135*t*
Festuca arundinacea, use to reduce amounts of PAHs in soil, 134–135
Fibrous root system
gross morphology, 200*f*
monocotyledons, 13
Fish, biochemical defense against toxicants, 83
Fixed-media reactors, biological treatment of wastes, 3
Flavobacterium sp.
degradation of pentachlorophenol, 179
use as inocula for bioremediation, 6
Flax, degradation of 2,4-D, 203*t*
Fluidized bed reactor, biological treatment of wastes, 3

Fluoranthene
distribution in soil organic matter in presence of *M. alba,* 89*t*
results of ^{14}C uptake experiment with *M. alba,* 89
uptake and accumulation in plants, 133
Fluorodifen, metabolism by glutathione conjugation, 185
Fluorodifen–GST (glutathione *S*-transferase) studies, 195–196
Free radicals
coupling reactions, mechanism, 102
from chlorinated phenols, 104
from 2,4-dichlorophenol oxidation, 106*f*
horseradish peroxidase-mediated binding of chlorophenols to humic acid, mechanism, 107
role in humification, 5
Freundlich isotherm to describe adsorption of naphthalene to roots, 136
Fulvic acid
association with ^{14}C-PAH, 82
[14]C
incorporation from ^{14}C-PAHs, 87
uptake experiment with *M. alba,* 89
complexes, use with surfactants, 61
density of stable free radicals, 107
mineralization
effect of bound surfactants, 68
LAE and LAS, 67*f*
root exudates, role, 58
Fumigation, partial sterilization effect, 46
Fungal enzymes, decomposition of anthropogenic pollutants, 19
Fungal populations
rhizosphere, 12
soil and compost mixes, 216,218*t*
Fungi
benefit from host plant, 94
benefit to plants, 16
detoxication of hazardous organic compounds, 11
mycorrhizal, 93–99
oxidoreductase activity, 103
pathogenic root-infecting, 93

Fungi—*Continued*
recycling of soil organic matter, 2
role in catabolism of organic substances, 12
Fusarium isolates, propanil aryl acylamidase activity, 170*t*

G

Galacturonic acids, role in fixed charge of cell walls, 77
Gas phase organics, analysis, 115
Gasoline tank leaks, cleanup by bioremediation, 6
Gautieria crispa, degradation of atrazine, 96
Gene transfer in bacteria, processes, 4
Genetic alteration of microorganisms, manipulation of mineralizing ability, 4
Gestrichum candidum, peroxidase activity, 103
Glucose
effect on PCP degradation, 43
use in PCP degradation study, 53–54*f*
Glutamate-treated soil, effect of PCP on bacterial population, 46,47*f*
γ-Glutamyltranspeptidase, in glutathione conjugate metabolism, 188
Glutathione conjugates, metabolism, 188,191,193
Glutathione conjugation, mechanism of detoxification, 185
Glutathione *S*-transferase, action, 184
Glycine
effect of PCP on degradation, 46,47*f*
effect on bacterial flora in percolated soil, 45*f*
Glycine max, enhanced mineralization of trichloroethylene, 202
Glycine-treated soil
effect of PCP on bacterial population, 44–46,48
microbiological processes, 46,47*f*
Gram-negative bacteria
in glycine-treated soil, 44

Gram-negative bacteria—*Continued*
  potential for propanil degradation,
    179–180
Grass
  growth as parameter in vegetative
    bioremediation, 148,149*t*
  mineralization of atrazine, 203*t*
  naphthalene volatilization, 19
  PAH degradation, 144
  PAH removal, 150
  roots, 13
  soil remediation, 134
  suitability for bioremediation, 148,151
Grass-enhanced bioremediation of clay
  soils, 142–157
Groundwater toluene levels, results,
  116–118
Gymnospermae, associations with
  endomycorrhizae, 17*t*

H

Halogenated aliphatics and aromatics,
  degradation by bacteria in the
  rhizosphere, 18*t*
Halogenated compounds, microbial
  degradation, 19,21
Hazardous compounds, degradation rate
  in rhizosphere, 102
Hazardous waste, amount generated in the
  United States, 3
Heavy metals, binding with mycorrhizal
  fungi, 96
Hebeloma crustuliniforme, catabolic
  enzymes, 94
Hebeloma cylindrosporum, degradation of
  chlorpropham, 97
Herbicide
  bioremediation in the rhizosphere, 19
  chemical stress to plants, 90
  degradation by mycorrhizal fungi,
    96–97
  enhanced degradation in rhizosphere,
    202,203*t*–204*t*

Herbicide—*Continued*
  nitrogen concentration, effect on
    degradation, 96
  rhizobacterial glutathione S-transferase,
    potential substrates, 185
Herbicide metabolism in higher plants, 84
Herbicide-tolerant plants, examples, 205
Heterotrophic processes, effect of PCP, 46
Hexazinone
  effect on ectomycorrhizae, 96
  immobilization by mycorrhizal fungi, 97
Holocellulose, degradation by
  mycorrhizae, 95
Horseradish peroxidase
  oxidative coupling of chlorophenols,
    104,105*t*
  wastewater treatment, 7
Host plant
  benefit from fungi, 94
  effects on detoxication, 11
Humic acid
  association with ^{14}C-PAH, 82
  binding to chlorinated phenols, 103
  ^{14}C
    incorporation from ^{14}C-PAHs, 87
    uptake experiment with *M. alba,* 89
  complexation, effect on mineralization
    of bound surfactants, 68
  covalent binding to chlorophenols, 109
  mineralization of LAE and LAS, 67*f*
  role of root exudates, 58
  stable free radicals, 107
  use of complexes with surfactants, 61
Humic substances
  binding to aromatic compounds, 7
  copolymerization of toxicants, 13
  modes of binding to xenobiotics, 109
  rhizosphere microorganisms, 13
Humification
  definition, 5
  role in detoxication, 86*f*
Humus
  binding to chlorinated phenols, 102–111
  effect on toxic pollutants, 5
  from unmineralized organic matter, 3

Hydrocarbon(s), degradation by bacteria
  in the rhizosphere, 18*t*
Hydrocarbon spills, cleanup by
  bioremediation, 6
Hydrolases, role as mediator in plant
  detoxication reactions, 84
Hydrolytic enzymes, metabolism of
  glutathione conjugates, 188,191,193
Hydroponic growth media,
  bioremediation studies, 91
4-Hydroxybenzoic acid, effect on growth
  of *A. lipoferum* strains, 32
2-Hydroxyphenylacetic acid, effect on
  growth of *A. lipoferum* strains, 32
Hymenoscyphus ericae
  degradation of 2,4-D, 96
  degradation of PCBs, 97
  metabolism of phytotoxic compounds, 96
Hyphae, position in mycorrhizal fungi, 94
Hyphal mats, produced by
ectomycorrhizae, 95
Hyphal network, formation by ectotrophic
  mycorrhizae, 94
Hysterangium gardneri, degradation of
  PCBs, 97
Hysterangium setchellii
  formation of hyphal mats, 95
  mat soils vs. nonmat soils, 95

I

Immobilization
  effect of plants, 11
  fate of chemical contaminants, 57
Immobilization vs. translocation, studies
  of phenol and bromacil, 114
In situ remediation
  bioremediation, 5–8
  cell-free enzymes, 7
  phytoremediation, 7
  soil remediation, 70
In situ rhizosphere degradation, influence
  of plants, 70–81

Industrial wastewaters
  biological treatment, 3
  use of bioreactors for treatment, 6
Industrially-impacted soil
  contamination by pollutant chemicals, 56
  sources of contamination, 57
Inoculation methods of bioremediation, 6
Insect(s)
  biochemical defense against toxicants, 83
  recycling of soil organic matter, 2
Insecticides
  accelerated degradation in the
    rhizosphere, 202
  chemical stress to plants, 90
Ion exchange, in plant cell wall, 77

K

K. R. Bluestem, seeded grass, 151
Kidman fine sandy loam, physiochemical
  properties, 126*t*
Kinetic parameters, mineralization of
  surfactants, 64*t*
Klebsiella planticola, alachlor
  metabolism, 186
Klebsiella spp., propanil aryl acylamidase
  activity, 169*t*
Knotweed, tolerance for herbicides, 205
Kochia sp.
  effect on atrazine, metolachlor, and
    trifluralin degradation, 207
  tolerance for herbicides, 205

L

Laccase
  fungi and actinomycetin, 103
  nonmotile *A. lipoferum,* 38
  oxidative coupling reactions, 103
  polymerization of chlorophenol, 107
Lagoons, biological treatment of wastes, 3
Laminarinase, in mat structures formed by
  ectomycorrhizae, 95

Landfarming, use in pollution control, 6
Leachability, effect of soil organic
matter, 13
Leaching
fate of chemical contaminants, 57
fate of naphthalene, mechanism, 124
Leguminous plants, association with plant
roots, 13
Lela clays, types, 134
Lespedeza cuneata, enhanced mineralization of trichloroethylene, 202
Leucine-treated soil, effect of PCP on
bacterial population, 46,47f
Lignin
degradation
by fungal enzymes, 19
by fungi, 6
by mycorrhizae, 95
compared to 2,4-D and atrazine
degradation, 96
mechanism, 75
role of mycorrhizal fungi, 95
metabolism to phenolic compounds, 28
similarity to PCB, 95
Lignocellulose, degradation by
mycorrhizae, 95
Lipophilic aromatic compounds,
biochemical transformations by
procaryotes, 86f
Lipophilic products, phenolic metabolism,
38,39

M

Mammals, biochemical defense against
toxicants, 83
Manganese, levels at mine spoil sites,
96
Mecoprop, carbon source for microbes in
wheat root, 203t
Medicago sativa, use to reduce amounts
of PAHs in soil, 134-135
Melanin, production by nonmotile *A.
lipoferum,* 38

Melilotus alba
experiment of microbial growth with soil
toxicant, 87
uptake studies, 88
Mesofauna, rhizosphere community, 12
Metabolic detoxication, plant protection, 82
Metabolic studies, releases by root
systems, 93
Metals
accumulation in plants, 7
contaminant, 3
electron transfer to metals via
phenolics, 29
Methane, monitoring in soil, 148
Methane monooxygenase, nonspecific
enzyme in cometabolism, 21
Methanogenic consortia, anaerobic
degradation of hazardous chemicals, 21
Methanotropic organisms, degradation of
TCE, 21
Methylcarbamoyloximes, transpiration
stream concentration factor values, 113
2-Methyl-4-chlorophenoxyacetic acid
(MCPA), carbon source for microbes in
wheat root, 203t
Metolachlor
contaminant from aquachemical
dealership, 205
enhanced degradation in rhizosphere
soil, 207
in pesticide-contaminated site, 206t
metabolism by glutathione conjugation,
185
occurrence in pesticide-contaminated
soil, 212
Metribuzin, metabolism by glutathione
conjugation, 185
Microbe-toxicant interactions, criteria
for establishing protective role, 87
Microbial activity
corn and soybean treatment, 64
in soil, 4-5
rhizosphere, 62-63
soil sampling for mineralization study,
61

Microbial biomass
    effect of rhizosphere, 62-63
    soil sampling for mineralization study, 61
Microbial degradation of toxicants
    catabolic pathways, 16
    in soils, use of vegetation, 11
Microbial enzymes that metabolize
    propanil, 164t
Microbial growth, effect on plants, 93
Microbial populations
    interactions in rhizosphere, 199,213,215t
    parameter in vegetative bioremediation,
        148,149t
    rhizosphere vs. nonrhizosphere soil, 103
    selective enrichment in rhizosphere, 184
    soil and compost mixes, 216-219,221
Microbial respiration, rhizosphere soil
    vs. nonvegetated soil, 207-208
Microbiological processes, soil treated
    with glycine, 46,47f
Microbiota, rhizosphere community, 12
Microcosm apparatus, 125f
Microfauna, rhizosphere community, 12
Microorganism(s)
    benefits from host plant, 85
    component of rhizosphere, 71
    effect of plants, 11
    influence on plant growth, 93
    nutrients, effect on interaction with
        PCP, 43-55
    role in soil bioremediation, 2-10
    soil minerals, movement into plants, 93
Microorganism-plant-chemical
    interactions in soil, 201-202
Mine spoil sites, levels of heavy metals, 96
Mineralization
    aerobic, 4
    bioremediation, 3
    effect of rhizosphere, 65
    field-collected soils with monocots and
        dicots, 73f
    interaction with volatilization, 130
    kinetics
        effect of rhizosphere, 56
        influence of soil organic matter, 56-69

Mineralization—Continued
    naphthalene, in microcosms, 128-129
    role in detoxication, 86f
    toxicants, resultant detoxication, 11
Mixed function oxidase system, induction
    compared with plant stimulation of
    microorganisms, 85
Mixed function oxidases, role as mediator
    in plant detoxication reactions, 84
Moisture, soil sampling for mineralization
    study, 61
Monocots, results of mineralization, 72
Monocotyledons
    2,4-D and 2,4,5-T degradation rates, 13
    roots and soil requirements, 13
Moraxella strain G, metabolism of
    3-chloroaniline, 163
Municipal wastewaters
    biological treatment, 3
    use of bioreactors for treatment, 6
Mycorrhizal fungi
    associations with vascular plants, 17t
    benefits to plants, 16
    effect on environmental pollutants,
        96-97
    enzymatic properties, 94
    growth and classification, 94
    metabolism, 94-95
    symbiotic relationship with plants, 5
    use as bioremediation agent, 93-99

N

Naphthalene
    adsorption to plant roots, 135-136
    ^{14}C uptake experiment with M. alba, 89
    common metabolites, 84f
    concentration in soil, 153,154t
    distribution in soil, 152
    distribution in soil organic matter in
        presence of M. alba, 89t
    grasses, effect on degradation, 144
    mineralization, 128-129

Naphthalene—*Continued*
physical properties and degradation
half-life, 124*t*
removal mechanism, 128
ring fissions, results, 86*f*
soil adsorption, effect on degradation, 76
vegetation, effect on mineralization, 19
volatilization and mineralization in
soil-grass microcosms, 123-131
volatilization potential, 127
Naphthalene-using bacteria in soil, 155
Natural soil
contamination by pollutant chemicals, 56
degradation, 56-69
sources of contamination, 57
Neutral sugars, effect on binding matrix, 68
Nitrate, from amino acids, 46
Nitrate-reducing bacteria, anaerobic
degradation of hazardous chemicals, 21
Nitrate-respiring bacteria, reductive
degradation of benzoate, 22*f*
Nitrification, inhibition by PCP, 46
Nitriles, cleanup by bioremediation, 6
Nitrite, from amino acids, 46
Nitrobacteraceae, description and
location, 15*t*
Nitrobenzenes, cleanup by
bioremediation, 6
Nitrogen fixation
dependence on O_2 concentration, 29
role in respiratory activity, 121
Nitrogen-fixing bacteria
association with plant roots, 13,76
symbiotic vs. nonsymbiotic, 28
Nitrogen-fixing nodules, formation, 13,16
Nitrogen-fixing prokaryotes, associations
with vascular plants, 14*t*
Nitrogenase activity
ATP requirement, 39
Azospirillum lipoferum, 35-38
Azospirillum spp., 29
stimulation by phenoxy acid herbicides,
40
Nitroreductases, role as mediator in plant
detoxication reactions, 84

Nitrous oxide (N_2O)
effect on acetylene-reduction activity, 39
redox potential changes, 39
Nitrous oxide (N_2O) reductase, inhibition
by acetylene, 29
Nonionic surfactants, biodegradation
studies, 56
Nonrhizosphere soil, nutrient
concentrations, 43
Nostoc calcicola, inhibition by propanil,
166
Nutrient(s)
degradation of PCP, 50-53
interaction between PCP and soil
microorganisms, 43-55
provided by mycorrhizal fungi, 94
Nutrient concentration
degradation of PCP, 53
rhizosphere, 12
Nutrient cycling, role of mycorrhizae, 95

O

Octanol-water partition coefficients
behavior of chemical classes, 113
polynuclear aromatic hydrocarbons, 143
Oldiodendron griseum
degradation of atrazine, 96
degradation of PCBs, 97
Orchid mycorrhizae, growth in axenic
culture, 94
Organic acids
effect on mineralization of pyrene, 139
organic rhizosphere exudates, 72
rhizosphere, 138
Organic matter
aerobic mineralization, 4
mineralization by microorganisms, 2
released from roots, 13
turnover in soil, 40
Organic pollutants
degradation of mycorrhizal fungi, 94
reasons for slow degradation, 4

Organic rhizosphere exudates, examples, 72
Organic solvents, contaminant, 3
Organophosphorus insecticides, inhibition
     of aryl amidase activity, 180
Orthoquinone
  relationship to catechol toxicity, 38
  tyrosinase-mediated oxidation of
     chlorophenols, 107
Oxidative coupling
  cause for intensification in
     rhizosphere, 103
  inorganic catalysts, 104
  reaction mechanism, 104
Oxidative pathways, polycyclic aromatic
     hydrocarbons, 20f
Oxidoreductase activity, bacteria and
     fungi, 103
Oxidoreductive enzymes
  catalysis of oxidative coupling
     reactions, 102
  dehalogenation effect, 103
  examples, 103
  role in humification, 5
Oxygen
  monitoring in soil, 148
  redox potential changes, 39
Oxygen partial pressure
  acetylene-reduction activity, 39
  growth of *A. lipoferum*, 38
  metabolism of caffeic acid, 39

P

Panicum virgatum
  effect on PAH degradation, 144
  use to reduce amounts of PAHs in soil,
     134-135
Parathion
  accelerated degradation in rhizosphere,
     184,202
  effect of plants on degradation rates, 123
  mineralization in bush bean rhizosphere,
     203t
Partial sterilization effect, 46

Pathogenic fungi
  cause of damage, 93
  root-infecting, 16
Pendimethalin, occurrence in
     pesticide-contaminated soil, 212
Penicillium frequentans, release of
     3,4-dichloroaniline from humic
     complexes, 163
Pentachloronitrobenzene, degradation by
     rhizosphere-competent fungus, 204t
Pentachlorophenol (PCP)
  concentration, effect on bacterial
     population, 44-46,48-50
  degradation by rhizosphere-competent
     fungus, 204t
  degradation in soil, 12
  effect of *Arthrobacter* sp., 179
  effect on bacterial flora
     glycine-treated soil, 48
     percolated soils, 44-48
  grasses, effect on degradation, 144
  nutrients
     effect on degradation, 50-53
     effect on interaction with soil
        microorganisms, 43-55
  soluble vs. insoluble form, 46
  uses, 43
Pentachlorophenol-degrading bacteria
  increase in numbers in soil, 48,49f,50
  nutrient concentration and degradability,
     effect on growth, 50-54f
  preparation in soil, 44
Perchloroethylene (PCE), degradation by
     anaerobic microorganisms, 4
Percolated soils
  bacterial flora, 48
  PCP treatment, 48
Perennial plants, roots, 13
Peroxidase
  activity, bacteria and fungi, 103
  mat structures formed by
     ectomycorrhizae, 95
  mediator in plant detoxication
     reactions, 84
  oxidative coupling reactions, 103

Pesticide(s)
  adsorption to plant roots, 133
  effect in agricultural soils, 56
  effect on soil microorganisms, 43
  fate in the rhizosphere, 199,201
  metabolism to phenolic compounds, 28
  phytoremediation, 7
  rhizosphere degradation studies, 71
Pesticide-contaminated soil
  characteristics that impede plant
    growth, 211
  compost, effect on plant growth and
    microbial establishment, 210-222
  concentrations of pesticides, 212
  effect on vegetation, 201
  remediation by direct land application,
    210-211
  soil-sampling procedure, 211
Pesticide-degrading microorganisms
  increase in soil with repeated treatment
    of pesticide, 50
  nutrient utilization, 43
Pesticide wastes, biological degradation
  in root zone of soil, 199-209
Petroleum products, contaminants, 3
Petroleum-related wastes, generation of
  phenolic compounds, 28
Petroleum sludge, disposal, 132
Phanerochaete chrysosporium
  bioremediation agent, 95
  degradation of xenobiotics, 97
  metabolism of PAHs, 19
  role in lignin degradation, 95
Phenacrycete chryosporium, propanil aryl
  acylamidase activity, 170*t*
Phenanthrene
  [14]C uptake experiment with *M. alba,*
    89
  concentration in soil, 153,154*t*
  distribution in soil organic matter in
    presence of *M. alba,* 89*t*
  grasses, effect on degradation, 144
  use as experimental toxicant, 87
Phenol
  catabolism by *Azospirillum,* 29

Phenol—*Continued*
  concentration results in saturated zone
    at several sampling wells, 117*t*
  degradation mechanism, 113-114
  degradation results by plant
    transpiration, 112
  evaporation rate, 120
  partition coefficient, 117
Phenolic compounds
  aerobic degradation, 29
  catabolism by *Azospirillum,* 28-29
  effect on *A. lipoferum,* 28-42
  fate in soil, 40
  microbial carbon source, 40
  organic rhizosphere exudates, 72
  redox potential changes, 39
  redox transformation, 38
  UV spectra after reaction with *A.
    lipoferum* cells or cell-free extract, 37*t*
Phenolic exudates from stimulated plants, 72
Phenolic redox couples, role in microbial
  adaptation to rice rhizosphere, 40
Phenoxy acid herbicides, stimulation of
  nitrogenase activity, 40
Phenylalanine-treated soil, effect of PCP
  on bacterial population, 46,47*f*
Phenylureas, transpiration stream
  concentration factor values, 113
Phosphatase, in mat structures formed by
  ectomycorrhizae, 95
Photodegradation, fate of chemical
  contaminants, 57
Photosynthate, inputs by roots, 13
Phthalate esters, reversal of toxic
  effects by soil microorganisms, 201
Physical plant factors in the
  rhizosphere, 75-76
Phytoremediation
  definition, 2,7
  stimulation of degradation rates, 2
Phytotoxic compounds, metabolism by
  mycorrhizal fungi, 96
Picloram, effect on ectomycorrhizae, 96
Pinus taeda, enhanced mineralization of
  trichloroethylene, 202

Plant(s)
  benefits for remediation, 211
  benefits from microorganisms, 85
  biological defenses against toxicants,
    84-85
  categories among species, 71-72
  coevolution with rhizosphere
    microorganisms, 85,87
  component of rhizosphere, 71
  effect on microbial growth in soil, 137
  effect on toxicant degradation, 12
  growth in pesticide-contaminated soil,
    213,214f,215t
  mechanisms of removing pollutants, 5
  protection from chemical injury by
    rhizosphere microbial community, 201
  protection from heavy metal toxicity by
    mycorrhizal fungi, 96
  remediation, future research, 140
  response to chemical stress in soil, 85
  role in dissipation of PAHs, 132
  selection of rhizosphere microbial
    communities, 202
  survival in presence of toxicants, role
    of microorganisms, 90
  uptake and metabolism of pollutants, 133
  uptake of PAHs, 143-144
  use in biodegradation study, 60
Plant-assisted bioremediation, 113
Plant degradation, availability of
    xenobiotics, 79
Plant growth, measurement, 115
Plant-induced rhizosphere, kinetic
    parameters for mineralization of
    surfactants in soils, 64t
Plant-microbe interactions
  development of rhizosphere community, 13
  evidence for coevolution, 85,87
  in rhizosphere, 201
  remediation of chemically contaminated
    soils, 11-12
Plant-microbe-toxicant interactions
  bioremediation, 11,90-91
  coevolution hypothesis, 90
  in the rhizosphere, 85

Plant root(s)
  adsorption of naphthalene, 135-136
  association with bacteria, 13,16
  enhancement of microbial degradation in
    soil, 21
  functions, 12-13
  release of organic chemicals into soil, 68
  sink for hazardous organic compounds,
    133
Plant root exudation
  effect of xenobiotics, 201
  interaction with ^{14}C surfactants, 68
Plant tissue microcosm, amount of
    radiolabeled carbon, 128t
Plant uptake, role in translocation,
    conjugation, and transformation of
    xenobiotics, 77
Plasmid transfer, effect on bacteria, 4
Plasticizers, cleanup by bioremediation, 6
Poisoned microcosm, amount of
    radiolabeled carbon, 128t
Pollutants
  absorption and metabolism by plants, 7
  effect of rhizosphere, 2
  mechanisms of removal by plants, 5
  phytoremediation, 7
  uptake and metabolism in plants, 133
Polyaromatic hydrocarbons (PAHs)
  adsorption to plant roots, 135-136
  degradation in greenhouse experiment,
    132-141
  dissipation in rhizosphere soil, 134
  dissipation in root zone, 132-141
  removal by grasses, 134
  uptake and accumulation in plants,
    133,135,136t
Polychlorinated biphenyls (PCBs)
  degradation by anaerobic
    microorganisms, 4
  degradation by mycorrhizal fungi, 97
  degradation by *P. chrysosporium,* 95
  metabolism by fungi, 93
Polycyclic aromatic hydrocarbons (PAHs)
  bioremediation in the rhizosphere, 19
  decomposition by fungal enzymes, 19

Polycyclic aromatic hydrocarbons
(PAHs)—*Continued*
humification in the rhizosphere, 88-89
metabolism by fungi, 19
oxidative pathways for mineralization, 20*f*
rhizosphere microbes, effect on
association with humic and fulvic
fractions, 90
role of vegetation, 131
sources and toxicity, 123
uptake by *M. alba*, 87
Polygalactic acids, in organic rhizosphere
exudates, 72
Polygonum sp., tolerance for herbicides,
205
Polymerization of toxicants, resultant
detoxication, 11,86*f*
Polynuclear aromatic hydrocarbons (PAHs)
concentration and distribution in soil,
152-156
degradation, enhancement by prairie
grasses, 143
degradation and polymerization in
plants, 143-144
grass, effect on degradation, 144
grass-enhanced degradation in clay
soils, 142-157
partitioning in clay soils, 142
soil type, effect on degradation, 155
uptake by plants, 143-144
Polyphenolic compounds
alternative terminal respiratory electron
acceptors under low oxygen, 28
degradation in catabolic pathway, 19
intermediary products in metabolism of
molecules with aromatic rings, 28
part of root exudates, 28
respiration by bacteria, 28-42
Polysaccharides
effect on binding matrix, 68
in organic rhizosphere exudates, 72
Prairie buffalo grass, description, 148
Prairie grasses, enhancement of PAH
degradation, 143
Prokaryotes, metabolism of PAHs, 19

Prokaryotic oxidative pathways,
mineralization of polycyclic aromatic
hydrocarbons, 22*f*
Propachlor
biodegradation pathways, 194
metabolism by glutathione conjugation,
185
Propanil
anaerobic hydrolysis, 170
anaerobic metabolism, 163
application to rice soils, 180
degradation by Gram-negative bacteria,
179-180
degradation by rice rhizosphere
bacteria, 167-168,171-174*f*
dissipation by rice rhizosphere
microflora, 160
dissipation in soil, 163
effect on plants, 166
herbicidal chemistry, 161,162*f*
hydrolysis by *Fusarium* spp., 170
metabolism
and mineralization, 168
plants, 161,162*f*
pure cultures, 176-178
rhizosphere microflora, 160-183
whole-cell suspensions, 170
microorganisms that can hydrolyze it,
163,164*t*
recovery from rhizosphere suspensions,
175*t*
studies on ring-labeled compound,
166-167
toxicity of metabolites, 179
Propanil aryl acylamidase activity
of fungi, 170
of rhizobacteria, 168-170
of rice rhizosphere enrichment cultures,
176*t*
Propionic acid, product of propanil
metabolism, 161
Protease, in mat structures formed by
ectomycorrhizae, 95
Proteins, degradation by cell-free
enzymes, 7

Protozoa, recycling of soil organic
  matter, 2
Psalliota arvense, tyrosinase activity, 103
Pseudomonadaceae strains, alachlor
  glutathione *S*-transferase, 185
Pseudomonas, aerobic degradation of
  phenolics, 29
Pseudomonas apacia, alachlor
  metabolism, 186
Pseudomonas cepacia
  decontamination of 2,4,5-T, 179
  propanil aryl acylamidase activity, 169*t*
  use as inocula for bioremediation, 6
Pseudomonas fluorescens
  alachlor metabolism, 186,187*f*
  propanil aryl acylamidase activity, 169*t*
Pseudomonas multivarans, metabolism of
  3-chloroaniline, 163
Pseudomonas purida, propanil aryl
  acylamidase activity, 169*t*
Pseudomonas putida
  alachlor metabolism, 186,187*f*
  use of 3,4-dichloroaniline as
    substrate, 163
Pseudomonas spp.
  adaptation to high root exudate levels, 13
  peroxidase activity, 103
  rhizosphere competence, 197
Pseudomonas strain G, metabolism of
  3-chloroaniline, 163,165*f*
Pseudomondaceae, description and
  location, 15*t*
Pteridophyta, associations with
  nitrogen-fixing prokaryotes, 14*t*
Pyrene
  adsorption to plant roots, 136
  dissipation in contaminated soil,
    137-138
  distribution in soil organic matter in
    presence of *M. alba,* 89*t*
  mineralization, 138-139
  resistance to degradation, 134
  uptake and accumulation in plants,
    133,135,136*t*
  vegetative remediation, 139-140

Q

Quinoline, effect of soil adsorption on
  degradation, 76
Quinone coenzymes, oxidation–reduction
  changes of polyphenolic compounds, 29

R

Radiigera atrogleba, degradation of
  PCBs, 97
Radiolabeled carbon, amount in poisoned,
  unvegetated, vegetated, and plant
  tissue microcosms, 128*t*
Radiolabeled organic compounds,
  cumulative volatilization in vegetated
  and unvegetated microcosms, 129*f*
Radionuclides, contaminant, 3
Redox potential
  determination, 31
  effect of caffeic acid, 39
  of medium with *A. lipoferum* and caffeic
    acid, 36*f*
  under electron acceptor-limiting
    conditions, 36*f*
Remediation, global importance, 70
Respiration rate
  effect of toluene and phenol withdrawal,
    121
  measurement, 116
Respiratory activity, role of nitrogen
  fixation, 121
Rhisosphere soil, nutrient concentrations,
  43
Rhizobacteria
  GST pathway of metabolism and
    detoxification, 196
  metabolism of alachlor with
    1-chloro–2,4-dinitrobenzene as
    substrate, 193–194
  plant growth promotion, 197
  propanil aryl acylamidase activity,
    168–170

Rhizobacteria cell-free extracts, alachlor-γ-glutamyltranspeptidase activity, 191t
Rhizobia, association with plant roots, 13
Rhizobiaceae, description and location, 15t
Rhizobium
degradation of aromatic and haloaromatic compounds, 28
nitrogen fixation, 28
propanil aryl acylamidase activity, 169t
Rhizoctonia praticola, laccase activity, 103
Rhizoctonia solani, metabolism of PAHs, 19
Rhizodeposition, definition and mechanism, 72
Rhizopogon vinicolor, degradation of 2,4-D, 96
Rhizopogon vulgaris, degradation of atrazine, 96
Rhizosphere
bacterial GST-mediated detoxification, site, 196
definition, 12,71,93,143,184
degradation of pesticides, 7
degradation of toxicants, 11–23
enhanced oxidative coupling, 103
gross morphology of fibrous root system, 200f
humification of PAHs, 88–89
increased microbial numbers after herbicide treatment, 207
mechanism of decreased mineralization, 65–68
microbial activity, stimulation by plants, 133
microbial biomass and activity, 56,62–63
microbial populations, determination, 212–213
mineralization of ^{14}C chemicals, 65
organic acids, 138
oxygen status, 16
physical dimensions and microbial activity, 12
phytoremediation, 7
plant–microbe interaction, 201

Rhizosphere—*Continued*
relationship to xenobiotic degradation, 71
symposium on applications to bioremediation technology, 12
transformation and degradation of pollutants, 2
Rhizosphere bacteria
fluorodifen–GST activity in cell-free extracts, 196t
glutathione *S*-transferase activity, 184–198
metabolism, effect on propanil dissipation in rice-production soils, 179
populations, effect of propanil, 171t
Rhizosphere degradation
branching pattern, 76
influence of plants, 70–81
root:shoot ratios, 76
volume of soil encountered by a root, 76
Rhizosphere enrichment
CDNB–GST activity, 193–194
effect of propanil, 171t
propanil metabolism, 168
Rhizosphere microbial communities
coevolution with plants, 85,87
copolymerization of toxicants, 13
humic substances, 13
microbial biodegradation studies, 56–69
pesticide-contaminated vs. uncontaminated vs. compost-treated soil, 213,215t
plant defense against toxic substances in soils, 82–92
role in protecting plants from chemical injury, 201
selection by plants, 201
Rhizosphere microbiota
degradation of naturally occurring toxicants, 16,19,21
effects on detoxication, 11
Rhizosphere microflora, metabolism of propanil, 160–183
Rhizosphere mineralization chamber, 59f
Rhizosphere soils
bacterial population, 13
vs. nonrhizosphere soil, microbial activity, 134

Rice, degradation of benthiocarb, 203*t*
Rice rhizosphere bacteria, propanil
  degradation, 167
Root(s)
  detoxication of hazardous organic
    compounds, 11
  sink for hazardous organic compounds,
    133
Root cell wall characteristics, role in
  rhizosphere degradation, 70
Root density, correlation to total
  microbial biomass, 113
Root epidermis, position relative to
  hyphae, 94
Root exudates
  examples, 58
  microbial growth, 93
  microbial nutrients, 43
  rate of chemical detoxication, 85
  removal and trap by plant microcosm, 73*f*
  rhizosphere, 13,70,72–75
Root-infecting fungi, types, 93–94
Root morphology, factors, 13
Root:shoot ratios, role in rhizosphere
  degradation, 76
Root systems, release of sugars and amino
  acids, 93
Russula nigricans, tyrosinase activity, 103

S

Sampling strategy, experiments on
  vegetative remediation, 146
Saprophytic fungi, 16,94
Schizachyrium scoparius, effect on PAH
  degradation, 144
Secretion and sequestration, defense
  against toxicants, 83
Serratia plymutica, alachlor metabolism,
  186
Serratia spp., propanil aryl acylamidase
  activity, 169*t*
Sewage sludges, effect in agricultural
  soils, 56–57

Simazine, effect on plant root exudation,
  201
Soil
  common bacteria, 15*t*
  component of rhizosphere, 71
  effect of plants, 11
  microorganism–plant–chemical
    interactions, 201–202
Soil aeration
  effect on biodegradation, 71
  parameter in vegetative bioremediation,
    148,149*t*
Soil analysis, methods, 115
Soil bacterial populations, effect of
  propanil, 171*t*
Soil bioremediation
  goal, 3
  role of microorganisms, 2–10
  typical pattern, 137
  use of *Azospirillum,* 28
  use of vegetation, 11
Soil dehydrogenase activity,
  determination, 213
Soil ecosystem, plant root system, 93
Soil fungi, types, 16
Soil gas monitoring, 148
Soil–grass microcosms, volatilization
  and mineralization of naphthalene,
  123–131
Soil microbial activity, 4–5
Soil microorganisms
  ability to detoxicate organic compounds,
    11
  binding of xenobiotics to soil, 102
  nutrients, effect on interaction with
    PCP, 43–55
  plant survival, 90
  protecting plants from chemical injury,
    201
Soil organic matter
  effects of pesticide degradation, 53
  influence on mineralization kinetics,
    56–69
  isolation, 90
  recycling by soil organisms, 2

Soil physicochemical characteristics, parameter in vegetative bioremediation, 148,149t
Soil properties, effects on detoxication, 11
Soil remediation study, statistical design, 145–147,155
Soil remediation technologies, examples, 70
Soil sampling, procedure, 148
Solvent treatment of soil, partial sterilization effect, 46
Sorghastrum nutans, effect on PAH degradation, 144
Sorghum vulgare, use to reduce amounts of PAHs in soil, 134–135
Sorptive immobilization, mechanism affecting fate of naphthalene, 124
Soybean
  effect of trichloroethylene, 202
  effect on atrazine, metolachlor, and trifluralin degradation, 207
  Glycine max, use in biodegradation study, 60
  use in biodegradation study, 56
Soybean treatment, effect on microbial activity, 64
St. Augustine grass, description, 150
Stern layer, model for selectively adsorbed ions, 77
Streptomyces spp.
  laccase and peroxidase activity, 103
Sudan grass, biomass of roots and shoots after growing for 24 weeks, 135t
Sugars, in organic rhizosphere exudates, 72
Suillus spp., degradation of chlorpropham, 97
Sulfate-reducing bacteria, anaerobic degradation of hazardous chemicals, 21
Superfund sites, cleanup costs, 3
Surfactants
  chemical stress to plants, 90
  contaminant in sewage sludge, 57
  effects of plants on degradation rates, 123
Switch grass, biomass of roots and shoots after growing for 24 weeks, 135t

Symbiotic relationship
  host plant and mycorrhizal fungi, 94
  nitrogen-fixing bacteria and plant roots, 76
  plants and mycorrhiza, 5
Systematic randomized sampling, experiments on vegetative remediation, 146

T

Tannins, metabolism to phenolic compounds, 28
Tap root system, dicotyledons, 13
Tetrachloroethylene (PCE)
  microbial degradation, 21
  reductive dechlorination, 23f
Texas Bluebonnet, use in bioremediation, 151
Thiocarbamates, metabolism by glutathione conjugation, 185
Toluene
  bioremediation in the rhizosphere, 19
  concentration results in saturated zone at several sampling wells, 117t
  degradation results by plant transpiration, 112
  partition coefficient, 117
Toluene dioxygenase, nonspecific enzyme in cometabolism, 21
Toxic pollutants, effects of humus, 5
Toxicant(s)
  aerobic vs. anaerobic metabolism, 16–21
  biological defenses
    animals and bacteria, 83
    plants, 84–85
  degradation in the rhizosphere, 11–23
  naturally occurring examples, 16
Toxicant interactions, degradation by microbiota, 16,19,21
Toxicant sorption, effect of soil organic matter, 13
Toxicology, fundamental precept, 83
Trametes versicolor
  laccase activity, 103

Trametes versicolor—Continued
  use in polymerization of chlorophenol,
    107
Transduction, method of gene transfer, 4
Transformation
  bioremediation, 3
  method of gene transfer, 4
  of xenobiotics, plant adsorption, 77
Translocation
  mechanism affecting fate of
    naphthalene, 124
  vs. immobilization, studies of phenol
    and bromacil, 114
  xenobiotics, plant adsorption, 77
Transpiration
  mechanism associated with microbial
    degradation, 114
  volatile compound movement through
    plants, 112
  vs. evaporation, effect on phenol
    concentration, 117
Trees, roots, 13
Triazines, metabolism by glutathione
  conjugation, 185
Trichloroethane, production of methane
  and chloride in saturated zone, 117
Trichloroethylene (TCE)
  accelerated degradation in rhizosphere,
    184
  degradation by root-associated
    microorganisms, 7
  degradation in rhizosphere soil, 134
  effect of plants on degradation, 113,123
  microbial degradation, 21,202
  production of methane and chloride in
    saturated zone, 117
2,4,5-Trichlorophenol
  coupling of free radicals, 106*f*,
    107–108
  dehalogenation due to binding to humic
    acid, 107
2,4,5-Trichlorophenoxyacetic acid
  (2,4,5-T), mineralization in soil
    exposed to dicot and monocot root
    exudates, 74*f*

Trichoderma harzaninum, propanil aryl
  acylamidase activity, 170*t*
Trickling filter, biological treatment of
  wastes, 3
Trifluralin
  contaminant from aquachemical
    dealership, 205
  enhanced degradation in rhizosphere
    soil, 207
  pesticide-contaminated site, 206*t*,212
Turf grasses, short-term benefits to
  contaminated soil, 144
Tyrosinase
  fungi and actinomycetin, 103
  oxidation of chlorophenols, mechanism,
    107
  oxidative coupling reactions, 103
  polymerization of chlorophenol, 107

U

Union Carbide
  contaminated sites, polynuclear
    aromatic hydrocarbons, 143
  Seadrift Plant, contamination history
    and description, 145
Unvegetated microcosm, amount of
  radiolabeled carbon, 128*t*

V

Vanillic acid, effect on growth of *A.
  lipoferum* strains, 32
Variability, experiments on vegetative
  remediation, 146
Vascular plants
  associations with mycorrhizae, 17*t*
  associations with nitrogen-fixing
    prokaryotes, 14*t*
Vegetated microcosm, amount of
  radiolabeled carbon, 128*t*
Vegetation
  advantages for soil remediation, 21

Vegetation—*Continued*
amounts of PAHs in soil, 134–135
bioremediation, 132–133
degradation of PAHs, 132
prerequisite for use in remediation, 135
problems of establishment in pesticide-
  contaminated sites, 201
Vegetative bioremediation
advantages, 140
experimental design of remedial
  performance evaluation, 145–146
fundamental premise, 139
parameters measured in performance
  evaluation, 148,149*t*
Verde Klein grass, description, 150
Vesicular–arbuscular mycorrhizae, 94
Vinyl chloride, accumulation under
  anaerobic conditions, 21
Vitamin B$_{12}$, reductive dechlorination of
  TCE, 23*f*
Volatilization
interaction with mineralization, 130
mechanism affecting fate of
  naphthalene, 124
naphthalene in microcosms, 128
promotion by alfalfa plants and
  associated microorganisms, 112–122

W

Waste chemicals
acceleration of mineralization through
  vegetation, 21
bioremediation in the rhizosphere, 19
Waste ponds, biological treatment of
  wastes, 3

Wastewater treatment, use of plants, 7
Water-percolated soil, bacterial flora
  and PCP treatment, 48
Weeping love grass, description, 150
Wheat, tolerance to phenoxy acids, 203*t*
Wheat straw, effects of plants on
  degradation rates, 123
White-rot fungus
degradation of xenobiotics, 97
lignin degradation, 95
soil inoculum, 6–7
Winecup, use in bioremediation, 151
Woodlot soil, biodegradation study, 56,58

X

Xanthomonas spp., propanil aryl
  acylamidase activity, 169*t*
Xenobiotic compounds
binding to humic substances, 109
in soil, 40,102
mineralization, 4
plant root exudation, 201
Xenobiotic uptake, role in rhizosphere
  degradation, 70
Xylanase, in mat structures formed by
  ectomycorrhizae, 95

Z

Zinc, levels at mine spoil sites, 96
Zoysia grass, description, 150
Zylenes, bioremediation in rhizosphere, 19

Production: Beth Harder & Charlotte McNaughton
Indexing: Scott Stoogenke & Colleen P. Stamm
Acquisition: Anne Wilson
Cover design: Amy Hayes

Printed and bound by Maple Press, York, PA

Highlights from ACS Books

Good Laboratory Practice Standards: Applications for Field and Laboratory Studies
Edited by Willa Y. Garner, Maureen S. Barge, and James P. Ussary
ACS Professional Reference Book; 572 pp; clothbound ISBN 0–8412–2192–8

Silent Spring Revisited
Edited by Gino J. Marco, Robert M. Hollingworth, and William Durham
214 pp; clothbound ISBN 0–8412–0980–4; paperback ISBN 0–8412–0981–2

The Microkinetics of Heterogeneous Catalysis
By James A. Dumesic, Dale F. Rudd, Luis M. Aparicio, James E. Rekoske, and Andrés A. Treviño
ACS Professional Reference Book; 316 pp; clothbound ISBN 0–8412–2214–2

Helping Your Child Learn Science
By Nancy Paulu with Margery Martin; Illustrated by Margaret Scott
58 pp; paperback ISBN 0–8412–2626–1

Handbook of Chemical Property Estimation Methods
By Warren J. Lyman, William F. Reehl, and David H. Rosenblatt
960 pp; clothbound ISBN 0–8412–1761–0

Understanding Chemical Patents: A Guide for the Inventor
By John T. Maynard and Howard M. Peters
184 pp; clothbound ISBN 0–8412–1997–4; paperback ISBN 0–8412–1998–2

Spectroscopy of Polymers
By Jack L. Koenig
ACS Professional Reference Book; 328 pp;
clothbound ISBN 0–8412–1904–4; paperback ISBN 0–8412–1924–9

Harnessing Biotechnology for the 21st Century
Edited by Michael R. Ladisch and Arindam Bose
Conference Proceedings Series; 612 pp;
clothbound ISBN 0–8412–2477–3

From Caveman to Chemist: Circumstances and Achievements
By Hugh W. Salzberg
300 pp; clothbound ISBN 0–8412–1786–6; paperback ISBN 0–8412–1787–4

The Green Flame: Surviving Government Secrecy
By Andrew Dequasie
300 pp; clothbound ISBN 0–8412–1857–9

For further information and a free catalog of ACS books, contact:
American Chemical Society
Distribution Office, Department 225
1155 16th Street, NW, Washington, DC 20036
Telephone 800–227–5558

Bestsellers from ACS Books

The ACS Style Guide: A Manual for Authors and Editors
Edited by Janet S. Dodd
264 pp; clothbound ISBN 0–8412–0917–0; paperback ISBN 0–8412–0943–X

The Basics of Technical Communicating
By B. Edward Cain
ACS Professional Reference Book; 198 pp;
clothbound ISBN 0–8412–1451–4; paperback ISBN 0–8412–1452–2

Chemical Activities (student and teacher editions)
By Christie L. Borgford and Lee R. Summerlin
330 pp; spiralbound ISBN 0–8412–1417–4; teacher ed. ISBN 0–8412–1416–6

Chemical Demonstrations: A Sourcebook for Teachers,
Volumes 1 and 2, Second Edition
Volume 1 by Lee R. Summerlin and James L. Ealy, Jr.;
Vol. 1, 198 pp; spiralbound ISBN 0–8412–1481–6;
Volume 2 by Lee R. Summerlin, Christie L. Borgford, and Julie B. Ealy
Vol. 2, 234 pp; spiralbound ISBN 0–8412–1535–9

Chemistry and Crime: From Sherlock Holmes to Today's Courtroom
Edited by Samuel M. Gerber
135 pp; clothbound ISBN 0–8412–0784–4; paperback ISBN 0–8412–0785–2

Writing the Laboratory Notebook
By Howard M. Kanare
145 pp; clothbound ISBN 0–8412–0906–5; paperback ISBN 0–8412–0933–2

Developing a Chemical Hygiene Plan
By Jay A. Young, Warren K. Kingsley, and George H. Wahl, Jr.
paperback ISBN 0–8412–1876–5

Introduction to Microwave Sample Preparation: Theory and Practice
Edited by H. M. Kingston and Lois B. Jassie
263 pp; clothbound ISBN 0–8412–1450–6

Principles of Environmental Sampling
Edited by Lawrence H. Keith
ACS Professional Reference Book; 458 pp;
clothbound ISBN 0–8412–1173–6; paperback ISBN 0–8412–1437–9

Biotechnology and Materials Science: Chemistry for the Future
Edited by Mary L. Good (Jacqueline K. Barton, Associate Editor)
135 pp; clothbound ISBN 0–8412–1472–7; paperback ISBN 0–8412–1473–5

For further information and a free catalog of ACS books, contact:
American Chemical Society
Distribution Office, Department 225
1155 16th Street, NW, Washington, DC 20036
Telephone 800–227–5558